THE THEORY AND APPLICATIONS OF HARMONIC INTEGRALS

THE THEORY AND APPLICATIONS OF HARMONIC INTEGRALS

BY

W. V. D. HODGE, M.A., F.R.S.

Lowndean Professor of Astronomy and Geometry,
and Fellow of Pembroke College,
Cambridge

With a foreword by Sir Michael Atiyah, F.R.S.

CAMBRIDGE UNIVERSITY PRESS

Cambridge

New York New Rochelle

Melbourne Sydney

CAMBRIDGE UNIVERSITY PRESS
Cambridge, New York, Melbourne, Madrid, Cape Town, Singapore, São Paulo, Delhi

Cambridge University Press
The Edinburgh Building, Cambridge CB2 8RU, UK

Published in the United States of America by Cambridge University Press, New York

www.cambridge.org
Information on this title: www.cambridge.org/9780521358811

First published 1941
Reissued with Foreword in the Cambridge Mathematical Library series 1989
Re-issued in this digitally printed version 2008

A catalogue record for this publication is available from the British Library

ISBN 978-0-521-35881-1 paperback

CONTENTS

FOREWORD

Hodge's book on Harmonic Integrals is one of the great landmarks of twentieth-century mathematics. Both in detail and in general outlook it set the stage for the global approach which has dominated geometry ever since. Hodge's work immediately attracted the attention of the leading figures of the time, including Lefschetz and Weyl who recognized its importance. The intervening decades have only served to reinforce their views, and Hodge theory continues to occupy a central position in contemporary research.

It is the fate of the pioneer to be the victim of his own success. Succeeding generations embellish and polish the first primitive steps so that future students will come to see the work as 'almost obvious'. Such is the back-handed tribute which posterity pays to its great innovators. For this reason the modern reader will find the technical presentation in Hodge's book rather cumbersome and pedestrian, but the classics are there to take us back to the birth of a subject, to show us the genesis of ideas as they appeared at the time. They are not meant to be vehicles of contemporary instruction.

When Hodge started his research in the early thirties he had before him the works of Picard and Lefschetz on algebraic integrals and their periods, together with the great body of classical algebraic geometry of the Italian school. His first breakthrough was his proof, following ideas of Lefschetz, that a (non-zero) double integral of the first kind (i.e. a holomorphic 2-form) on an algebraic surface has non-trivial periods. Ironically Lefschetz at first refused to accept the proof and campaigned fiercely against it. Eventually he was converted and became Hodge's strongest supporter.

Hodge soon realized that the main difference between curves and surfaces was that in the latter case holomorphic forms and their conjugates do not exhaust the cohomology.

The reason, as we now know, is the existence of harmonic forms of type (1,1) and Hodge's ideas gradually evolved in this direction. In due course Hodge proved his index or signature theorem which gave a topological interpretation to the geometric genus p_g (the dimension of the space of holomorphic 2-forms). Hodge's formula can be written as

$$b_2^+ = 2p_g + 1$$

where $b_2 = b_2^+ + b_2^-$ is the decomposition of the second Betti number according to the signature of the intersection matrix of 2-cycles. This signature formula was the first great success of Hodge theory, and it made a big impact.

The striking thing about the Hodge signature theorem is that the result is intrinsic to the geometry and topology of the surface but its proof requires the introduction of an auxiliary tool, namely a Kähler metric. This, in the end, is the justification for the introduction of real analysis into algebraic geometry.

More generally the Hodge numbers $h^{p,q}$ (the dimension of the space $H^{p,q}$ of harmonic forms of type (p,q)) are invariants of algebraic varieties, independent of the choice of Kähler metric. Only with the introduction of sheaf cohomology in the fifties were the $h^{p,q}$ given an intrinsic definition, namely

$$h^{p,q}(X) = \dim H^q(X, \Omega^p)$$

where Ω^p is the sheaf of holomorphic p-forms on X.

With the enormous explosion of algebraic geometry since the introduction of sheaf theory, Hodge theory has continued to play a key role. The relation between the (p,q) decomposition and the integral lattice of homology gives an elaborate structure to the period relations. This has been extensively studied by Griffiths and subsequently by Deligne who has coined the term 'Hodge structure' to describe these period relations.

Although Hodge's motivation came from algebraic geometry and this is the area in which it has had its deepest and most striking applications, the theory does of course apply in the first instance to general Riemannian manifolds. As such it is part of global differential geometry and there have been notable applications in this context. The famous Bochner 'vanishing theorems' provide a direct link between positivity conditions on curvature and vanishing of Betti numbers (the extension by Kodaira to Kähler manifolds is also important in algebraic geometry). Also harmonic forms have proved useful in connection with Lie groups. This was already recognized, for compact groups, by Hodge and figures as the last chapter of his book. Subsequently applications to discrete subgroups of non-compact Lie groups have been made by applying Hodge theory to compact locally symmetric spaces.

Also, in a Riemannian context, Hodge theory has a parallel in the analysis of the Dirac equation and the study of harmonic spinors. These were studied extensively by Atiyah and Singer in connection with the index theory of elliptic differential operators and they are also important in present-day physics.

In physical terms Hodge theory has been reinterpreted by Witten in the framework of super-symmetric quantum mechanics. The Hodge Laplacian is the Hamiltonian of the quantum mechanical system. This use of Hodge theory in physics appears to be quite far-reaching and it extends, at least formally, to the context of certain quantum field theories. According to Witten these should be viewed as Hodge theory on appropriate infinite-dimensional manifolds.

In a different but related direction Hodge theory has had a remarkable generalization in the study of the Yang–Mills equations. These are non-linear analogues of the Hodge equations and they have been exploited with spectacular success by Donaldson for the study of 4-dimensional manifolds. Moreover, Donaldson imitates Hodge by using Kähler metrics on algebraic surfaces to define invariants which are then shown to have a purely algebro-geometric significance. This work of Donaldson can thus be viewed as a non-linear generalization of the Hodge signature theorem. In fact the signature of the algebraic surface (or rather b_2^+) enters into the dimension formula for the Donaldson moduli spaces of solutions of the Yang–Mills equations precisely through their linearization.

This very brief review hardly does justice to the scope of Hodge theory at the present time but it does indicate its richness and variety. The connections with theoretical physics would have particularly pleased Hodge since he had been much influenced by Maxwell's equations and was for many years a colleague of Dirac.

Michael Atiyah
1988

REFERENCES

M. F. Atiyah and I. M. Singer, The index of elliptic operators I, *Ann. of Math.* **87** (1968), 484–530.

S. K. Donaldson, An application of gauge theory to 4-dimensional topology, *J. Diff. Geom.* **18** (1983), 279–315.

P. Griffiths and J. Harris, *Principles of Algebraic Geometry*, Wiley, New York, 1978.

E. Witten, Supersymmetry and Morse theory, *J. Diff. Geom.* **17** (1982), 661–92.

PREFACE

The subject of this book is the study of certain integrals defined in a type of space which is of importance in various branches of mathematics. The space is locally the space of classical Riemannian geometry, and in the large it is an orientable manifold. By considering simultaneously the local and general properties of the space, we are led to the study of a class of integrals in the space to which the name harmonic integral has been given.

In its original form the book formed a section of an essay for which the Adams Prize of 1936 was awarded, but since then it has been revised and entirely re-written, and the subject-matter has been enlarged by the addition of a chapter dealing with the application of the harmonic integrals to the theory of continuous groups.

The first chapter is concerned with the geometry of the space in which the integrals are defined. The properties which are required for the work of later chapters fall into two classes, those relating to the differential geometry of the space, and those which are topological properties. Both these subjects are treated at length in a number of standard works, and there seems no reason to add to the existing literature. I have therefore contented myself with a brief survey of the bare essentials, but I hope that I have said enough on these topics to enable the reader who is unacquainted with Riemannian geometry or topology to understand the later chapters. The second chapter deals with the properties of integrals on a manifold, and in it I give a proof of de Rham's theorem on the existence of an integral with assigned periods, while the third chapter introduces harmonic integrals and contains a proof of the fundamental existence theorem for these integrals.

The remainder of the book is concerned with the applications of the theory of harmonic integrals to other branches of mathematics. It is clear that applications of our theory will be possible in any field of mathematical research in which a Riemannian manifold plays a part. But when the differential geometry of the manifold has special properties we are able to go much further with our theory than in the general case. I have not attempted to invent any manifolds in which the conditions are particularly favourable for the development of the properties of harmonic integrals, and I have confined my attention to manifolds which arise naturally in two important branches of mathematics.

In Chapter IV I consider the properties of harmonic integrals in the Riemannian of an algebraic variety. It is necessary to introduce a metric which is irrelevant in the classical theory of varieties, but the greater part of the chapter is devoted to deducing, from the properties of harmonic integrals, invariants of the manifold which do not depend on the metric. Most of the results obtained belong to the transcendental theory of varieties, but a few geometrical results can be deduced by the methods which we employ. But, while it is possible to explain the results belonging to the transcendental theory without requiring much knowledge of the theory of algebraic varieties on the part of the reader, a considerable knowledge of algebraic geometry is required in order to understand the significance of the geometrical applications of our theory. Since these applications are at present somewhat isolated, and can only be regarded as preliminary, there does not seem to be sufficient justification for prefacing this part of the chapter with a lengthy account of the algebraic geometry of varieties. I have therefore confined my account of the geometrical applications of harmonic integrals to two paragraphs [§§ 51, 52] which really form an appendix to the chapter, and are merely intended to direct the attention of geometers to the possibility of further investigations.

Chapter V considers the application of the theory of harmonic integrals to certain problems in the theory on continuous groups. The reader will require some slight knowledge of group theory in this chapter, but, following the precedent of Chapter I, I have begun the chapter with a brief summary of the results which will be used. The chapter shows that our theory provides an alternative method of considering the invariant integrals introduced by Cartan in the topological theory of groups. In a number of important cases the results obtained coincide exactly with those found by Cartan. The chapter concludes with the determination of the Betti numbers of the group manifolds associated with the four main classes of simple groups. In this I follow closely the work of Brauer and Weyl, though in places it is modified by the use of properties of harmonic integrals.

In the earlier stages of preparing this book I had the advantage of much useful criticism from Dr J. H. C. Whitehead, of Balliol College, Oxford, but after the outbreak of war it was not possible for me to continue receiving the benefit of his advice. Prof. T. A. A. Broadbent, of the Royal Naval College, Greenwich, has read the manuscript, and has helped greatly in reading the proofs. I wish to express my thanks to both of these gentlemen for their great assistance, and to the staff of the Cambridge University Press for the care which they have taken in the printing of this book.

W. V. D. H.

Pembroke College, Cambridge
September 1940

Chapter I

RIEMANNIAN MANIFOLDS

1·1. **Introduction.** The theory of harmonic integrals has its origin in an attempt to generalise the well-known existence theorem of Riemann for the everywhere finite integrals on a Riemann surface. In making the generalisation, the first necessity is to determine the nature of the n-dimensional space which is to play the part of the Riemann surface. The space which we obtain is called an n-dimensional Riemannian manifold. A Riemannian manifold of two dimensions is not, however, the same as a Riemann surface, and, as an introduction to the ideas with which we shall deal, we shall first consider the difference between the two concepts. These considerations will lead up to, and may help to elucidate, the formal definition of Riemannian manifolds of n dimensions which will be given later.

Let us construct in the usual way the Riemann surface for the algebraic equation, over the field of complex numbers,

$$F(z, w) = 0,$$

which defines w as an algebraic function of the variable z. This Riemann surface is a closed, orientable (i.e. two-sided) surface on which we introduce certain local coordinate systems, which we call allowable coordinate systems. In the neighbourhood of a place on the surface for which $z = a$ we take as one allowable coordinate system (σ, τ), where

$$z-a = t = \sigma+i\tau, \quad \text{or} \quad z-a = t^n = (\sigma+i\tau)^n,$$

according as the place is the origin of a linear branch, or a branch of order n. If z is infinite at the place, we replace $z-a$ by z^{-1}. Then (x_1, x_2) are allowable coordinates in the neighbourhood of the place if

$$y = x_1+ix_2 = f(t)$$

is an analytic function of t which is simple (*schlicht*) in the neighbourhood of the origin. All the allowable coordinate systems are obtained in this way, and we may call y an allowable complex parameter in the neighbourhood of the place.

If (x_1, x_2) and (x'_1, x'_2) are two allowable coordinate systems in the neighbourhood of a place, x'_1 and x'_2 are analytic functions of (x_1, x_2), which satisfy certain equations (the Cauchy-Riemann equations), and there exists a relation

$$(dx'_1)^2 + (dx'_2)^2 = \lambda[(dx_1)^2 + (dx_2)^2]$$

between the differentials at a point, where λ is a positive analytic function. Thus by means of the allowable coordinate systems we define a local geometry on the Riemann surface which is known as *conformal geometry*. In this geometry, length has no significance, but angles can be defined.

If we make a birational transformation of the fundamental algebraic equation, we obtain a new equation and corresponding to it a new Riemann surface. We denote the original Riemann surface by R and the new one by R'. There is a (1-1) continuous correspondence between the points of R and R', that is, R and R' are homeomorphic. The homeomorphism has, however, some special properties. If P and P' are corresponding points of R and R', and (x_1, x_2) and (x'_1, x'_2) are allowable local coordinates valid in their neighbourhoods, the equations giving the homeomorphism in the neighbourhoods of P and P' are

$$x'_i = f'_i(x_1, x_2) \qquad (i = 1,\ 2),$$
$$x_i = f_i(x'_1, x'_2) \qquad (i = 1,\ 2),$$

where the functions are analytic. This homeomorphism is therefore an *analytic homeomorphism*. Moreover,

$$x'_1 + ix'_2 = X'(x_1 + ix_2),$$
$$x_1 + ix_2 = X(x'_1 + ix'_2),$$

where the functions are analytic, and hence

$$(dx_1)^2 + (dx_2)^2 = \mu[(dx'_1)^2 + (dx'_2)^2],$$

where μ is a positive function. Thus the conformal properties of R and R' are preserved in the homeomorphism. The homeomorphism is therefore a conformal representation of the one surface on the other. Conversely, if two Riemann surfaces are conformally representable on one another by an analytic homeomorphism, the algebraic equations to which they correspond are birationally equivalent. The surfaces can therefore be regarded as equivalent.

1·2. The features of a Riemann surface which we wish to emphasise are that it is a closed orientable surface carrying certain allowable coordinate systems which specify a local geometry, and that between equivalent Riemann surfaces there is a homeomorphism which relates the local geometries of the surfaces. A Riemannian manifold of two dimensions is also a closed orientable surface, but differs from a Riemann surface in the systems of coordinates which are allowable, and in the local geometry.

The allowable coordinate systems on a Riemannian manifold of two dimensions are characterised by the properties (*a*) that there is at least one allowable system valid in the neighbourhood of any point, and (*b*) that, if (x_1, x_2) are allowable coordinates in a neighbourhood N, and x_1' and x_2' are differentiable functions of (x_1, x_2) in N, the necessary and sufficient conditions that (x_1', x_2') are allowable coordinates in N are:

(i) the Jacobian $$\frac{\partial(x_1', x_2')}{\partial(x_1, x_2)}$$

is different from zero in N and

(ii) (x_1', x_2') do not assume the same set of values at two distinct points of N. If the functions x_1', x_2' have continuous derivatives of order u, where u is a positive integer or zero, they are said to be *of class u* (class ω if they are analytic); and if the equations of transformation relating any two allowable systems of coordinates are of class u, the manifold is said to be *of class u*.

While the local geometry of a Riemann surface is conformal geometry, the local geometry on a Riemannian manifold of two dimensions is *Riemannian geometry.* We associate with each allowable coordinate system (x_1, x_2) a positive definite quadratic differential form

$$E\,dx_1^2 + 2F\,dx_1\,dx_2 + G\,dx_2^2,$$

where E, F, G are functions of (x_1, x_2), and are of class $(u-1)$ if the manifold is of class u. If

$$E'dx_1'^2 + 2F'dx_1'dx_2' + G'dx_2'^2$$

is the differential form associated with another coordinate system (x_1', x_2') valid in the same neighbourhood, the coefficients of the two expressions are connected by the equations

$$E' = E\left(\frac{\partial x_1}{\partial x_1'}\right)^2 + 2F\frac{\partial x_1}{\partial x_1'}\frac{\partial x_2}{\partial x_1'} + G\left(\frac{\partial x_2}{\partial x_1'}\right)^2,$$

$$F' = E\frac{\partial x_1}{\partial x_1'}\frac{\partial x_1}{\partial x_2'} + F\left(\frac{\partial x_1}{\partial x_1'}\frac{\partial x_2}{\partial x_2'} + \frac{\partial x_1}{\partial x_2'}\frac{\partial x_2}{\partial x_1'}\right) + G\frac{\partial x_2}{\partial x_1'}\frac{\partial x_2}{\partial x_2'},$$

$$G' = E\left(\frac{\partial x_1}{\partial x_2'}\right)^2 + 2F\frac{\partial x_1}{\partial x_2'}\frac{\partial x_2}{\partial x_2'} + G\left(\frac{\partial x_2}{\partial x_2'}\right)^2.$$

The quadratic differential forms enable us to define in an invariant way the length of any arc on the manifold, as in elementary differential geometry.

Two Riemannian manifolds of two dimensions which are of class u are equivalent if there exists a (1-1) correspondence between their points which is given locally in terms of allowable coordinate systems by equations of class u, and is such that the lengths of corresponding arcs are the same. If two Riemannian manifolds of two dimensions are in (1-1) correspondence of class u, we can always choose allowable coordinate systems in the neighbourhoods of any pair of corresponding points so that in these neighbourhoods corresponding points have the same coordinates. The necessary and sufficient condition that the manifolds should be equivalent is that in these coordinate systems the coefficients of the quadratic differential forms on the two manifolds should be equal at corresponding points. If the manifolds are not equivalent, it may yet happen that in these coordinate systems the coefficients of the quadratic differential forms are proportional at corresponding points. In this case we say that the manifolds are *conformally related* by the homeomorphism.

1·3. There is clearly a considerable difference in character between a Riemann surface and a Riemannian manifold of two dimensions. It is therefore worth while showing how we can pass from the one to the other, and how we can obtain properties of a Riemann surface from a knowledge of the properties of a Riemannian manifold of two dimensions. Among the allowable systems of coordinates on a Riemannian manifold we can find a sub-set G with the property that the equations of transformation from one coordinate system of G to another are analytic. In G there is a sub-set G_1 for which the fundamental quadratic form is given by

$$\lambda\,(dx_1^2 + dx_2^2).$$

If (x_1, x_2) and (x_1', x_2') are two coordinate systems of G_1 valid in the same region, it is well known that

$$x_1' + ix_2' = f(x_1 \pm ix_2),$$

where the function is analytic. There is a sub-set G_2 of G_1 for which the upper sign holds. The closed orientable surface which is the Riemannian manifold, and which has G_2 as the set of allowable coordinate systems, is a Riemann surface. Riemannian manifolds of two dimensions which are conformally homeomorphic define in this way equivalent Riemannian surfaces. In a later chapter we shall see that, conversely, a Riemann surface defines, in a unique way, an infinite set of Riemannian manifolds of two dimensions, any two of which are conformally homeomorphic. Certain invariants, in particular those which we shall call harmonic integrals, of a Riemannian manifold are unaltered when we pass from one manifold to a conformally homeomorphic manifold, and therefore define invariants of a Riemann surface. While this method of obtaining invariants of a Riemann surface is extremely artificial, the generalisation of it enables us to obtain invariants of the Riemannian of an algebraic variety of any number of dimensions which have not as yet been obtained by other means. But there are also other fields in which we can apply the theory of Riemannian manifolds.

2·1. **Manifolds of class u.** We now define a Riemannian manifold of dimension n. These come within the category of manifolds of class u, as defined by Veblen and Whitehead [10]†, and the reader is referred to their tract for a more elaborate examination of the structure of such manifolds than we give here. We shall give only a brief description of their character.

In order to define any space, we begin with a set of undefined elements which we call *points*, and impose certain conditions on them. We postulate the existence of certain sub-sets, which we call *neighbourhoods*, and we suppose that every point lies in at least one neighbourhood. We call the neighbourhoods which contain a point the neighbourhoods of the point, and when we say that a property holds in the neighbourhood of a point we mean that there is at least one neighbourhood of the point in which the property holds.

The neighbourhoods with which we shall deal are also assumed to have the properties:

(*a*) if any two neighbourhoods N and N' have a point in common, there is a neighbourhood of this point which lies in both N and N';

(*b*) if P and Q are distinct points of the set, there exist neighbourhoods of P and Q which have no point in common.

A set of points and a set of neighbourhoods with these properties form a *space*. By putting further restrictions on the neighbourhoods we can determine different types of space.

† References given in this way relate to the list of references at the end of each chapter.

The space which we now define is called a *manifold*. We require first that the points of any neighbourhood shall be in (1-1) correspondence with the set N of points of the real number space $(x_1, \ldots, x_n)$ of n dimensions which satisfy the inequalities

$$|x_i| < \delta \qquad (i = 1, \ldots, n)$$

where δ is a given positive number. If $(\bar{x}_1, \ldots, \bar{x}_n)$ is any point of N, there exists a number δ_1 such that all the points given by

$$|x_i - \bar{x}_i| < \delta_1 \qquad (i = 1, \ldots, n)$$

lie in N, and for any such δ_1 we suppose that those points of the space which correspond to the points of N satisfying

$$|x_i - \bar{x}_i| < \delta_1 \qquad (i = 1, \ldots, n)$$

form a neighbourhood.

The number n is the same for all neighbourhoods, and is called the *dimension* of the manifold.

The correspondence between a neighbourhood of a manifold and the points of the number space given by

$$|x_i| < \delta \qquad (i = 1, \ldots, n)$$

defines a *coordinate system* in the neighbourhood. Since any point lies in several neighbourhoods, there exist several coordinate systems valid at a point. Any two which are valid at the same point are both valid in some neighbourhood of the point. Let $(x_1, \ldots, x_n)$ and $(x'_1, \ldots, x'_n)$ be two coordinate systems valid in a neighbourhood N. Then, at points of N there is a (1-1) correspondence between the two systems of coordinates which can be expressed by the equations

$$x_i = f_i(x'_1, \ldots, x'_n) \qquad (i = 1, \ldots, n),$$

$$x'_i = g_i(x_1, \ldots, x_n) \qquad (i = 1, \ldots, n).$$

Such a system of equations defines a transformation of coordinates. By saying that the manifold is of class u, we mean that the neighbourhoods are so restricted that the functions $f_i(x'_1, \ldots, x'_n)$, $g_i(x_1, \ldots, x_n)$, for $i = 1, \ldots, n$, are of class u, and that, when $u>0$, the equations of transformation are *regular* in N, that is, that the Jacobians

$$\left|\frac{\partial f_i}{\partial x'_j}\right|, \quad \left|\frac{\partial g_i}{\partial x_j}\right|$$

are different from zero at all points of N. The ideas of *limit point*, etc. on the manifold can be defined in terms of these coordinate systems.

It is convenient to admit further allowable coordinate systems in a neighbourhood N. Let $y_1, \ldots, y_n$ be real functions of $(x_1, \ldots, x_n)$ of class u, where $(x_1, \ldots, x_n)$ are coordinates of the type already defined, and suppose that the functions are defined at all points of N and satisfy the conditions:

(i) $\left|\frac{\partial y_i}{\partial x_j}\right|$ is different from zero in N (when $u>0$); and

(ii) there exists no pair of points x and x' in N for which

$$y_i(x) = y_i(x') \qquad (i=1, \ldots, n).$$

Each point of N can then be identified by the set of values of the functions $y_1, \ldots, y_n$ at the point. We shall therefore admit $(y_1, \ldots, y_n)$ as allowable coordinates in N.

We now impose two conditions on the neighbourhoods which restrict the nature of the manifold as a whole. First, the manifold must be *finite*; that is, there must be a finite number of neighbourhoods $N_1, \ldots, N_r$ whose sum entirely covers the manifold. Secondly, the manifold must be *connected*; that is, the neighbourhoods $N_1, \ldots, N_r$ cannot be divided into sets, each non-vacuous, $N_{i_1}, \ldots, N_{i_s}$ and $N_{i_{s+1}}, \ldots, N_{i_r}$, which have no point in common.

2·2. A manifold, as we have defined it, can be represented in a simple manner as a locus in Euclidean space. In an n-dimensional number space $(x_1, \ldots, x_n)$, any set of points in (1-1) continuous correspondence with the points interior to a sphere of the space is called a *simplicial region* of the space. We consider a set of points in the Euclidean space $(X_1, \ldots, X_N)$ which (i) lies entirely in a finite region of the space; (ii) forms a connected set; (iii) has the property that those points of the set which lie within a distance δ of any given point P of the set are in (1-1) correspondence of class u with the points of a simplicial region N in $(x_1, \ldots, x_n)$, and are given by equations

$$X_i = f_i(x_1, \ldots, x_n) \qquad (i = 1, \ldots, N),$$

where the functions are of class u, and where the matrix

$$\left(\frac{\partial X_i}{\partial x_j}\right)$$

is of rank n at all points of N. Such a set of points is said to form a *locus of class u* in $(X_1, \ldots, X_N)$. We now propose to show that any manifold of class u can be represented as a locus of class u_1 in a Euclidean space, where u_1 is any finite integer not exceeding u.

To prove this, we consider the neighbourhoods $N_1, \ldots, N_r$ whose sum covers the manifold. Let $(x_1^j, \ldots, x_n^j)$ be a coordinate system in N_j obtained by representing N_j on

$$|x_i^j| < \delta \qquad (i = 1, \ldots, n)$$

in the number space $(x_1^j, \ldots, x_n^j)$. We define $2nr$ functions y_i^j, z_i^j of the points of the manifold as follows. The functions y_i^j, z_i^j are zero at all points not in N_j. At points of N_j,

$$y_i^j = [(x_i^j)^2 - \delta^2]^s,$$

$$z_i^j = x_i^j y_i^j.$$

If $s > u_1$, these functions, regarded as functions of allowable coordinates valid in the neighbourhood of any point of the manifold, are functions of class u_1. A point P of the manifold lies in at least one neighbourhood N_j, and for this j, y_i^j is not zero at P. Let P and Q be two distinct points of the manifold. If P lies in N_j and Q does not lie in N_j, then

$$y_i^j(P) \neq 0, \quad y_i^j(Q) = 0.$$

If both P and Q lie in N_j, and if

$$y_i^j(P) = y_i^j(Q), \quad z_i^j(P) = z_i^j(Q),$$

then

$$x_i^j(P) = x_i^j(Q).$$

Thus we can always find i, j so that either

$$y_i^j(P) \neq y_i^j(Q),$$

or

$$z_i^j(P) \neq z_i^j(Q).$$

Now let $(X_1, \ldots, X_{2nr})$ be coordinates in a Euclidean space of $2nr$ dimensions. Consider the locus defined by

$$X_{(j-1)n+i} = y_i^j,$$

$$X_{(r+j-1)n+i} = z_i^j,$$

for $i = 1, \ldots, n; j = 1, \ldots, r$, where the argument of the functions on the right-hand sides is a point P which ranges over the manifold. This locus is in (1-1) correspondence with the manifold. It is easily verified that it is a locus of class u_1, as defined above. Conversely, we may show that a locus of class u_1 is a manifold of class u_1, the neighbourhoods being defined in the obvious way.

A representation of a manifold of class u as a locus of class u_1 in a Euclidean space will be referred to as a Euclidean representation, and will be found useful later in proving some general theorems. We observe that if u is finite we may take $u_1 = u$, but if $u = \omega$ this is not possible (but see Whitney [13]).

2·3. At this stage we may interpolate a remark concerning the class number u. We shall not be particularly interested in proving our results under a minimum number of conditions imposed on the manifold, and for the applications which we shall make it will only be necessary to suppose that u is sufficiently large for the operations which we perform. In fact, it will appear that every result which we establish will be valid if $u > 6$, and many will be valid for smaller values of u. We shall often leave as an exercise to the reader the problem of determining the smallest value of u for which a given result is true, and we shall always assume, without explicit statement, that u is large enough for our purpose.

3·1. The Riemannian metric. We now make the hypothesis that the class number u of a Riemannian manifold is greater than zero. We may therefore differentiate the equations of transformation connecting two coordinate systems $(x_1, \ldots, x_n)$ and $(x'_1, \ldots, x'_n)$ which are valid in the same neighbourhood. We denote the differential of x_i by dx^i†. The equations connecting the differentials dx^i, dx'^i at a point are

$$dx'^i = \sum_{h=1}^{n} \frac{\partial x'_i}{\partial x_h} dx^h,$$

$$dx^i = \sum_{h=1}^{n} \frac{\partial x_i}{\partial x'_h} dx'^h.$$

In writing equations which involve functions to which one or more indices are attached, we shall make use of the summation convention, by which we agree to sum over the possible values of the indices with respect to each index which appears twice, once at the top and once at the bottom. Thus we shall write the above equations as

$$dx'^i = \frac{\partial x'_i}{\partial x_h} dx^h,$$

$$dx^i = \frac{\partial x_i}{\partial x'_h} dx'^h.$$

Let $(x_1, \ldots, x_n)$ be any coordinate system, valid in a neighbourhood N of the manifold, and let dx^i be the differential

† To adhere strictly to the conventions which we adopt we should write the coordinates of a point as $(x^1, \ldots, x^n)$, as is done in several works. But, since we shall in the earlier part of our work have to consider quadratic functions of the coordinates, it is more convenient typographically to write the coordinates with indices at the foot. But when we consider differentials of coordinates it is necessary to adopt the conventional rules more strictly, and write the indices at the top. In a derivative $\partial x_i/\partial x_j'$ appearing in the equation of transformation of differentials the suffix i is to be regarded as an upper index and j as a lower index.

of x_i. We associate with each point P of N a positive definite quadratic differential form in the differentials dx^i

$$g_{ij}dx^i dx^j \qquad (g_{ij}=g_{ji}),$$

where we suppose that the forms associated with the different points of N are such that the coefficients g_{ij} are functions of $(x_1, \ldots, x_n)$ of class $u-1$. There will be a form associated with P for each coordinate system valid in a neighbourhood of P. If $(x_1, \ldots, x_n)$ and $(x'_1, \ldots, x'_n)$ are two such coordinate systems, and

$$g_{ij}dx^i dx^j, \quad g'_{ij}dx'^i dx'^j$$

are the corresponding forms at P, we require the coefficients of the forms to be connected by the relations

$$g'_{ij} = g_{hk}\frac{\partial x_h}{\partial x'_i}\frac{\partial x_k}{\partial x'_j}.$$

Then it is only necessary to give the form at P for one coordinate system, since the others may be obtained uniquely by this rule. If, at every point P of a manifold of class u, we are given a series of quadratic differential forms, one for each coordinate system valid in a neighbourhood of P, which satisfy the properties which we have described, we say that the manifold carries a Riemannian metric. If

$$x_i = x_i(t) \qquad (t_0 \leqslant t \leqslant t_1)$$

is any arc on the manifold, we define its length to be

$$\int_{t_0}^{t_1}\left\{g_{ij}\frac{dx_i}{dt}\frac{dx_j}{dt}\right\}^{\frac{1}{2}} dt.$$

The length of the arc does not depend on the coordinate system used. Two manifolds of class u, each carrying a Riemannian metric, are regarded as equivalent if (i) there is a (1-1) correspondence between their points which is given locally in terms of allowable coordinate systems by equations of class u, and (ii) corresponding arcs are of the same length.

3·2. In the applications which can be made of the theory of harmonic integrals there are two ways in which the metric can arise. In the first case, the metric is an integral part of the problem, and is completely defined by it. In the other case, the metric is not defined as part of the data, and in attaching a metric to the manifold a certain amount of latitude is possible. The choice may be completely free, or it may be limited but still allow a certain amount of freedom. If we are given a Euclidean representation of the manifold in the space $(X_1, \ldots, X_N)$ by equations

$$X_i = f_i(x_1, \ldots, x_n) \qquad (i = 1, \ldots, N),$$

where the functions are of class u, we shall sometimes find it convenient to introduce the metric by means of the Euclidean distance element

$$ds^2 = dX_1^2 + \ldots + dX_N^2.$$

By "the metric defined by the Euclidean distance element" we mean the metric given by the equation

$$g_{ij}dx^i dx^j = \sum_{h=1}^{N} \left(\frac{\partial f_h}{\partial x_i} dx^i\right)^2.$$

If the metric is once given, and we then construct a Euclidean representation of the manifold, it is of course clear that the given metric will in general be different from that defined by the Euclidean distance element. The problem of finding a Euclidean representation with the property that the metric defined by the Euclidean distance element coincides with an assigned metric is one which has as yet been solved only in special cases.

A case which is of some importance is that in which the manifold is defined as an analytic locus in Euclidean space. In this case, the coefficients of the form giving the metric defined by the Euclidean distance element are analytic.

4·1. **Orientation.** A manifold of class u, carrying a Riemannian metric, may be called a Riemannian manifold, and the term Riemannian manifold is used in this sense by M. Morse [7]. In the theory which we are going to develop one other property of the space is essential, and we shall define a Riemannian manifold as a space satisfying the conditions of §§ 2 and 3, and the further condition that it is *orientable*. The property of orientability is defined as follows.

If $(x_1, \ldots, x_n)$ and $(x'_1, \ldots, x'_n)$ are two allowable coordinate systems on a manifold, both valid in a neighbourhood N, we have imposed the condition that the Jacobian

$$\left| \frac{\partial x_i}{\partial x'_j} \right|$$

should be different from zero at all points of N. Since this Jacobian is a continuous function of the points of N, it has the same sign at all points of N. If the sign is positive, we say that the coordinate systems $(x_1, \ldots, x_n)$ and $(x'_1, \ldots, x'_n)$ are "like" in N, and if it is negative we say that they are "unlike". Clearly, if $(x_1, \ldots, x_n)$ and $(x'_1, \ldots, x'_n)$ are unlike, $(x_1, \ldots, x_n)$ and $(x'_1, \ldots, x'_{n-1}, -x'_n)$ are like.

Now consider the set of all the allowable coordinate systems on the manifold. Let there be a sub-set of these with the properties:

(i) every point of the manifold lies in the domain of at least one of the systems of the sub-set;

(ii) if two systems of the sub-set are valid in the same neighbourhood N, they are like in N.

We then say that the manifold is *orientable*. An orientable manifold associated with a set of like coordinate systems is called an *oriented manifold*, and clearly one orientable manifold defines two oriented manifolds. If no sub-set of allowable coordinate systems satisfying (i) and (ii) exists, the manifold is not orientable.

Let $N_1, \ldots, N_r$ be a set of neighbourhoods covering an oriented manifold. We can find a coordinate system $(x^j_1, \ldots, x^j_n)$ valid in

N_j which is like the systems of the sub-set used to orient the manifold, and then, if N_h and N_k have points in common,

$$\left|\frac{\partial x_i^h}{\partial x_j^k}\right|$$

is positive at these points. Conversely, if coordinate systems $(x_1^j, \ldots, x_n^j)$ can be found in the neighbourhoods N_j $(j = 1, \ldots, r)$ such that

$$\left|\frac{\partial x_i^h}{\partial x_j^k}\right|$$

is positive at any point common to N_h and N_k, the manifold is orientable. Thus, to determine whether a manifold is orientable or not, we have only to consider coordinate systems valid in a finite number of neighbourhoods whose sum covers the manifold.

4·2. Consider some examples. If (θ, ϕ) are polar coordinates on a sphere in space of three dimensions, the manifold can be covered by the two neighbourhoods given by

$$0 \leqslant \theta < \tfrac{1}{2}\pi + \delta \qquad (N_1),$$

$$\tfrac{1}{2}\pi - \delta < \theta \leqslant \pi \qquad (N_2).$$

We may take as coordinates in N_1,

$$x_1^1 = \tan \tfrac{1}{2}\theta \cos \phi, \quad x_2^1 = -\tan \tfrac{1}{2}\theta \sin \phi,$$

and as coordinates in N_2,

$$x_1^2 = \cot \tfrac{1}{2}\theta \cos \phi, \quad x_2^2 = \cot \tfrac{1}{2}\theta \sin \phi.$$

Then at points common to N_1 and N_2,

$$\frac{\partial(x_1^1, x_2^1)}{\partial(x_1^2, x_2^2)} = \frac{\partial(x_1^1, x_2^1)}{\partial(\theta, \phi)} \Bigg/ \frac{\partial(x_1^2, x_2^2)}{\partial(\theta, \phi)} = \tan^4 \tfrac{1}{2}\theta > 0,$$

and hence the sphere is orientable.

On the other hand, if we consider the real projective plane (x, y, z), we can cover it by three neighbourhoods N_1, N_2, N_3, where N_1 is given by

$$|y/x| < 2, \quad |z/x| < 2; \quad x_1^1 = y/x, \; x_2^1 = z/x;$$

N_2 is given by

$$|z/y| < 2, \quad |x/y| < 2; \quad x_1^2 = z/y, \; x_2^2 = x/y;$$

N_3 is given by

$$|x/z| < 2, \quad |y/z| < 2; \quad x_1^3 = x/z, \; x_2^3 = y/z.$$

Each pair of these neighbourhoods has two regions in common; in one of these the Jacobian of transformation is positive and in the other it is negative. It follows easily from this that the projective plane is not an orientable manifold.

5. **Geometry of a Riemannian manifold.** We have now completed the definition of a Riemannian manifold. To sum up this definition, a space is a Riemannian manifold of class u if

(i) it is a manifold of class u;

(ii) it carries a Riemannian metric;

(iii) it is an orientable manifold.

In order to develop the theory of integrals, and, in particular, of harmonic integrals, on a Riemannian manifold, a knowledge of the geometry of the manifold is necessary. This geometry falls into two parts. The first consists of the local properties, and is, in fact, the study of the geometry defined by the metric, that is, Riemannian geometry. The second consists of the geometry in the large, that is, the combinatorial topology of the manifold. Both these topics are treated in several excellent textbooks, to which the reader may refer, but for convenience we give a brief résumé of the theories, in so far as our requirements demand, so that the reader may have the essential results before him. Complete proofs of each result stated would make the account unduly long, and since they are available in other places to which references are made, they are omitted from the following sections.

DIFFERENTIAL GEOMETRY

In this section we are concerned with the local geometry on a Riemannian manifold M. In practice, we find that we are only concerned with properties at a point P and at points in the neighbourhood of P. We may therefore confine our attention to a neighbourhood N, and without any loss of generality we may assume that the neighbourhoods in which any of the allowable coordinate systems which we introduce are valid contain N. The results obtained in this section will be found more fully treated in the works of Eisenhart [4] and Veblen [9], and elsewhere.

6·1. **Tensors and their algebra.** We have already spoken about differentials at a point. The differentials $dx^1, \ldots, dx^n$ are arbitrary and independent numbers. When a change of coordinate system $x \to x'$ is effected, they are replaced by $dx'^1, \ldots, dx'^n$, defined by the non-singular linear transformation

$$dx'^i = \frac{\partial x'_i}{\partial x_j} dx^j, \tag{1}$$

where the partial derivatives are evaluated at P. The object at P, which consists of the association of a set of numbers $(dx^1, \ldots, dx^n)$ with each allowable coordinate system at P, with the law of transformation (1) is called a *contravariant vector* at P. The numbers of the set used to define the vector in any coordinate system are the *components* of the vector in this coordinate system. The use of the differential symbol is irrelevant, and it is not at all necessary to think of the components of a contravariant vector as the differentials of a set of functions. We write the components of a contravariant vector as $(\xi^1, \ldots, \xi^n)$ and we denote it by its generic component ξ^i. The index is written at the top in order to distinguish contravariant vectors from covariant vectors which we are about to define, and the fact that the differentials of the coordinates are components of a contravariant vector is our reason for writing the differential of x_i as dx^i.

A *covariant vector* at P is defined as a object given by components $(\eta_1, \ldots, \eta_n)$ associated with each coordinate system, with the transformation law

$$\eta'_i = \frac{\partial x_j}{\partial x'_i} \eta_j,$$

the partial derivatives having their values at P. If ξ^i, η_i are, respectively, contravariant and covariant vectors at P, we have

$$\begin{aligned} \xi'^i \eta'_i &= \frac{\partial x'_i}{\partial x_h} \xi^h \frac{\partial x_k}{\partial x'_i} \eta_k \\ &= \delta^k_h \xi^h \eta_k \\ &= \xi^h \eta_h, \end{aligned}$$

where $\delta^k_h = 0$ if $h \neq k$, and $\delta^h_h = 1$. The object $\xi^i \eta_i$ is called a *scalar invariant*, since it is independent of the coordinate system.

6·2. The vectors which we have defined are special cases of a large class of geometric invariants called *tensors*, which we now define. A tensor is an object (associated with a point P) given by n^{p+q} components

$$T^{j_1 \ldots j_q}_{i_1 \ldots i_p} \qquad (i_r, j_s = 1, \ldots, n)$$

in the coordinate system $(x_1, \ldots, x_n)$, with the property that its components $T'^{j_1 \ldots j_q}_{i_1 \ldots i_p}$ in any other coordinate system $(x'_1, \ldots, x'_n)$ are given by

$$T'^{j_1 \ldots j_q}_{i_1 \ldots i_p} = \left| \frac{\partial x_i}{\partial x'_j} \right|^W T^{b_1 \ldots b_q}_{a_1 \ldots a_p} \frac{\partial x_{a_1}}{\partial x'_{i_1}} \cdots \frac{\partial x_{a_p}}{\partial x'_{i_p}} \frac{\partial x'_{j_1}}{\partial x_{b_1}} \cdots \frac{\partial x'_{j_q}}{\partial x_{b_q}}, \tag{2}$$

where W is an integer. The tensor is generally represented by its generic component. The lower indices are called *covariant indices*, and the upper indices *contravariant indices*. W is the *weight* of the tensor, and $(p+q)$ is its *rank*. If there are no contravariant indices we speak of the tensor as a covariant tensor, and if there are no covariant indices the tensor is said to be a contravariant tensor. If $W = 0$, the tensor is said to be *absolute*, and if $p = q = 0$ it is *scalar*. Two tensors with the same number of indices of each type, and with the same weight, are said to be *like*. If the components of a tensor are all zero in one system of coordinates, the law of transformation shows that they are zero in every system of coordinates, and the tensor is then called a *zero tensor*.

From the law of transformation we also see that if $T^{j_1 \dots j_q}_{i_1 \dots i_p}$ is symmetric or skew-symmetric in i_r, i_s (or in j_r, j_s), $T'^{j_1 \dots j_q}_{i_1 \dots i_p}$ is also symmetric or skew-symmetric in i_r, i_s (or in j_r, j_s). The most important case of this occurs when the tensor is a covariant tensor or a contravariant tensor, and when it is symmetric, or skew-symmetric, in each pair of indices. We then say that the tensor is a symmetric tensor, or a skew-symmetric tensor. From the definition of the Riemannian metric

$$g_{ij} dx^i dx^j$$

it follows that g_{ij} is an absolute symmetric covariant tensor. We call it the (covariant) metrical tensor. $g_{ij} dx^i dx^j$ is an absolute scalar.

6·3. Certain algebraic operations on tensors are permissible.

I. *Addition of tensors.* If $T^{j_1 \dots j_q}_{i_1 \dots i_p}$, $S^{j_1 \dots j_q}_{i_1 \dots i_p}$ are like tensors, at a point, a tensor which is like to each can be obtained by adding corresponding components in the same coordinate system. This follows at once from the linear nature of the law of transformation. The sum is denoted by

$$T^{j_1 \dots j_q}_{i_1 \dots i_p} + S^{j_1 \dots j_q}_{i_1 \dots i_p}.$$

II. *Multiplication of tensors.* $T^{j_1 \dots j_q}_{i_1 \dots i_p}$, $S^{j_1 \dots j_s}_{i_1 \dots r_t}$ are two tensors at a point of weights W_1 and W_2 respectively, given by their components in the coordinate system $(x_1, \dots, x_n)$. Let

$$R^{j_1 \dots j_{q+s}}_{i_1 \dots i_{p+r}} = T^{j_1 \dots j_q}_{i_1 \dots i_p} S^{j_{q+1} \dots j_{q+s}}_{i_{p+1} \dots i_{p+r}}.$$

We can define a tensor of weight $W_1 + W_2$ by saying that its components in the coordinate system $(x_1, \dots, x_n)$ are the numbers $R^{j_1\ldots j_{q+s}}_{i_1\ldots i_{p+r}}$ $(i_r, j_s = 1, \ldots, n)$. If we now determine its components in the coordinate system $(x'_1, \ldots, x'_n)$, we have

$$R'^{j_1\ldots j_{q+s}}_{i_1\ldots i_{p+r}} = \left|\frac{\partial x_i}{\partial x'_j}\right|^{W_1+W_2} T^{b_1\ldots b_q}_{a_1\ldots a_p} S^{b_{q+1}\ldots b_{q+s}}_{a_{p+1}\ldots a_{p+r}} \frac{\partial x_{a_1}}{\partial x'_{i_1}} \cdots \frac{\partial x_{a_{p+r}}}{\partial x'_{i_{p+r}}} \frac{\partial x'_{j_1}}{\partial x_{b_1}} \cdots \frac{\partial x'_{j_{q+s}}}{\partial x_{b_{q+s}}}$$

$$= T'^{j_1\ldots j_q}_{i_1\ldots i_p} S'^{j_{q+1}\ldots j_{q+s}}_{i_{p+1}\ldots i_{p+r}}.$$

Thus the tensor $R^{j_1\ldots j_{q+s}}_{i_1\ldots i_{p+r}}$ is determined uniquely by the tensors $T^{j_1\ldots j_q}_{i_1\ldots i_p}$ and $S^{j_1\ldots j_s}_{i_1\ldots i_r}$, independently of the coordinate system. It is called the *product* of the two tensors.

III. *Contraction of tensors*. Let $T^{j_1\ldots j_q}_{i_1\ldots i_p}$ be a tensor of weight W, given by its components in the coordinate system $(x_1, \ldots, x_n)$. Replace i_r, j_s by k, and apply the summation convention. Write

$$S^{j_1\ldots j_{q-1}}_{i_1\ldots i_{p-1}} = T^{j_1\ldots j_{s-1}\,k\,j_s\ldots j_{q-1}}_{i_1\ldots i_{r-1}\,k\,i_r\ldots i_{p-1}}.$$

We define a tensor $S^{j_1\ldots j_{q-1}}_{i_1\ldots i_{p-1}}$ of weight W by giving the components the above values in the coordinate system $(x_1, \ldots, x_n)$. For convenience we consider the case in which $r = p$ and $s = q$, but the following argument is valid in all cases. When we make a transformation of coordinates we obtain the new components

$$S'^{j_1\ldots j_{q-1}}_{i_1\ldots i_{p-1}} = \left|\frac{\partial x_i}{\partial x'_j}\right|^{W} S^{b_1\ldots b_{q-1}}_{a_1\ldots a_{p-1}} \frac{\partial x_{a_1}}{\partial x'_{i_1}} \cdots \frac{\partial x_{a_{p-1}}}{\partial x'_{i_{p-1}}} \frac{\partial x'_{j_1}}{\partial x_{b_1}} \cdots \frac{\partial x'_{j_{q-1}}}{\partial x_{b_{q-1}}}$$

$$= \left|\frac{\partial x_i}{\partial x'_j}\right|^{W} T^{b_1\ldots b_q}_{a_1\ldots a_p} \frac{\partial x_{a_1}}{\partial x'_{i_1}} \cdots \frac{\partial x_{a_{p-1}}}{\partial x'_{i_{p-1}}} \frac{\partial x_{a_p}}{\partial x'_k} \frac{\partial x'_{j_1}}{\partial x_{b_1}} \cdots \frac{\partial x'_{j_{q-1}}}{\partial x_{b_{q-1}}} \frac{\partial x'_k}{\partial x_{b_p}}$$

$$= T'^{j_1\ldots j_{q-1}k}_{i_1\ldots i_{p-1}k}.$$

The tensor $S^{j_1\ldots j_{q-1}}_{i_1\ldots i_{p-1}}$ is therefore defined uniquely by the process which we have described, from $T^{j_1\ldots j_q}_{i_1\ldots i_p}$. We describe the process by which we pass from $T^{j_1\ldots j_q}_{i_1\ldots i_p}$ to $S^{j_1\ldots j_{q-1}}_{i_1\ldots i_{p-1}}$ as *contraction*. Since we may give r and s any values, it follows that from $T^{j_1\ldots j_q}_{i_1\ldots i_p}$ we can obtain pq tensors by contraction. By repeating the process, further tensors may be obtained.

Given any finite number of tensors, by repeated application of the rules for the addition, multiplication and contraction of tensors we obtain other tensors. If in this way we arrive at a zero tensor we have a tensor equation connecting the tensors, which holds in all coordinate systems.

7·1. Numerical tensors. The metrical tensors. Certain particular tensors which are of frequent occurrence may be noted here.

(i) Let
$$\delta^{j_1 \ldots j_p}_{i_1 \ldots i_p}$$
be a number which is zero unless $i_1, \ldots, i_p$ and $j_1, \ldots, j_p$ are derangements of the same p distinct integers lying between 1 and n, and which is $+1$ when $i_1, \ldots, i_p$ is an even derangement of $j_1, \ldots, j_p$, and is -1 when it is an odd derangement. We define a tensor of weight zero at a point by taking its components in the coordinate system $(x_1, \ldots, x_n)$ to be

$$\Delta^{j_1 \ldots j_p}_{i_1 \ldots i_p} = \delta^{j_1 \ldots j_p}_{i_1 \ldots i_p}.$$

Its components $\Delta'^{j_1 \ldots j_p}_{i_1 \ldots i_p}$ in another coordinate system are

$$\begin{aligned}\Delta'^{j_1 \ldots j_p}_{i_1 \ldots i_p} &= \delta^{b_1 \ldots b_p}_{a_1 \ldots a_p} \frac{\partial x_{a_1}}{\partial x'_{i_1}} \cdots \frac{\partial x_{a_p}}{\partial x'_{i_p}} \frac{\partial x'_{j_1}}{\partial x_{b_1}} \cdots \frac{\partial x'_{j_p}}{\partial x_{b_p}} \\ &= \begin{vmatrix} \dfrac{\partial x'_{j_1}}{\partial x'_{i_1}} \cdots \dfrac{\partial x'_{j_1}}{\partial x'_{i_p}} \\ \vdots \\ \dfrac{\partial x'_{j_p}}{\partial x'_{i_1}} \cdots \dfrac{\partial x'_{j_p}}{\partial x'_{i_p}} \end{vmatrix} \\ &= \delta^{j_1 \ldots j_p}_{i_1 \ldots i_p}.\end{aligned}$$

Thus there is a unique tensor whose components in any coordinate system satisfy the equations

$$\Delta'^{j_1 \ldots j_p}_{i_1 \ldots i_p} = \delta^{j_1 \ldots j_p}_{i_1 \ldots i_p}.$$

We denote this tensor by

$$\delta^{j_1 \ldots j_p}_{i_1 \ldots i_p}.$$

(ii) By a similar argument we prove by direct calculation that there is a skew-symmetric tensor, which we denote by $\epsilon_{i_1\ldots i_n}$, of weight -1, whose components in any coordinate system satisfy the equations

$$\epsilon_{i_1\ldots i_n} = \delta^{1\ldots n}_{i_1\ldots i_n}.$$

(iii) Similarly, we show that there is a skew-symmetric tensor $\epsilon^{i_1\ldots i_n}$ of weight $+1$ whose components in any coordinate system satisfy the equations

$$\epsilon^{i_1\ldots i_n} = \delta^{i_1\ldots i_n}_{1\ldots n}.$$

A tensor of weight $+1$ is called a *tensor density.*

7·2. The coefficients of the differential form

$$g_{ij}\,dx^i\,dx^j$$

are components of an absolute symmetric covariant tensor, which we have called the covariant metrical tensor. In any coordinate system we can solve the equations

$$g^{ij}g_{jk} = \delta^i_k,$$

for the unknowns g^{ij}, since $|g_{ij}| \neq 0$, the quadratic form being definite. Clearly

$$g^{ij} = g^{ji}.$$

We define g^{ij} in this way for each coordinate system. We now define a tensor h^{ij} by the condition that in the particular coordinate system $(x_1, \ldots, x_n)$ we have

$$h^{ij} = g^{ij}.$$

From the invariance of tensor equations we have

$$h'^{ij}g'_{jk} = \delta^i_k$$

in any other coordinate system. Hence

$$h'^{ij} = g'^{ij}$$

in every coordinate system. We denote the tensor h^{ij} by g^{ij}, and call it the *contravariant metrical tensor*.

Finally, if
$$g = |g_{ij}|,$$
we have
$$g' = |g'_{ij}| = |g_{hk}| \left|\frac{\partial x_h}{\partial x'_i}\right|^2$$
$$= \left|\frac{\partial x_i}{\partial x'_j}\right|^2 g,$$

and hence g is a scalar of weight 2. Similarly, $\sqrt{g}$ is a tensor density.

8·1. **Parallel displacement.** The tensors which we have discussed are located at a definite point of our manifold, and form a vector space, which we call the *tangent space*, at the point. A (1-1) correspondence between the tensors at a point P and the tensors at a point P' is called a *parallelism* if

(i) absolute scalars at P correspond to equal absolute scalars at P';

(ii) when $S^{j_1 \ldots j_q}_{i_1 \ldots i_p}$, $T^{j_1 \ldots j_s}_{i_1 \ldots i_r}$, ... are tensors at P, and $S'^{j_1 \ldots j_q}_{i_1 \ldots i_p}$, $T'^{j_1 \ldots j_s}_{i_1 \ldots i_r}$, ... are the (like) corresponding tensors at P', then the tensor obtained by performing any series of operations of addition, multiplication, and contraction on $S^{j_1 \ldots j_q}_{i_1 \ldots i_p}$, $T^{j_1 \ldots j_s}_{i_1 \ldots i_r}$, ... corresponds to the tensor obtained by performing the same series of operations on $S'^{j_1 \ldots j_q}_{i_1 \ldots i_p}$, $T'^{j_1 \ldots j_s}_{i_1 \ldots i_r}$, ...

In differential geometry it is necessary to establish a parallelism between the tangent spaces at any two points P and P' of the manifold. It is not necessary that this parallelism should be unique, and we shall see that the parallelism which we introduce depends not only on P and P', but also on the choice of an arc on the manifold joining P and P'.

In a suitable coordinate system, let P be the point $(\bar{x}_1, \ldots, \bar{x}_n)$ and P' be $(\bar{x}_1 + \epsilon^1 t, \ldots, \bar{x}_n + \epsilon^n t)$. We shall be chiefly concerned with values of t which tend to zero, and in the equations which follow terms in t of the second or higher orders are of no importance. Let ξ^i be an absolute contravariant vector at P, and denote the parallel vector at P' by $\xi^i + d\xi^i$. Let us consider the case in which $d\xi^i$ is given by an equation of the form

$$d\xi^i = -L^i_{jk}\xi^j \epsilon^k t + O(t^2),$$

where L^i_{jk} depends only on P (in the given coordinate system). Neglecting terms of the second and higher order in t, and writing $\epsilon^i t = \delta x^i$, we say that the equation

$$d\xi^i + L^i_{jk}\xi^j \delta x^k = 0$$

defines the infinitesimal parallel displacement of the vector ξ^i at P corresponding to the displacement δx^i.

We now consider a change of coordinate system. In the following equations a partial derivative enclosed in brackets denotes the value of that derivative at P, and a partial derivative enclosed in brackets with a suffix denotes the value of the derivative at the point indicated by the suffix. Let P have coordinates $(\bar{x}'_1, \ldots, \bar{x}'_n)$ in the new coordinate system, and let P' have coordinates $(\tilde{x}'_1, \ldots, \tilde{x}'_n)$. Then

$$\begin{aligned}\tilde{x}'_i &= x'_i(\bar{x}' + \epsilon t)\\ &= \bar{x}'_i + \left(\frac{\partial x'_i}{\partial x_j}\right)\epsilon^j t + O(t^2)\\ &= \bar{x}'_i + \epsilon'^i t,\end{aligned}$$

where $$\epsilon'^i = \left(\frac{\partial x'_i}{\partial x_j}\right)\epsilon^j + O(t).$$

The new components of $\xi^i + d\xi^i$ are

$$\begin{aligned}\xi'^i + d\xi'^i &= (\xi^j + d\xi^j)\left(\frac{\partial x'_i}{\partial x_j}\right)_{P'}\\ &= (\xi^j + d\xi^j)\left[\left(\frac{\partial x'_i}{\partial x_j}\right) + \left(\frac{\partial^2 x'_i}{\partial x_j\,\partial x_k}\right)\epsilon^k t + O(t^2)\right]\\ &= \xi'^i + \left(\frac{\partial^2 x'_i}{\partial x_j\,\partial x_k}\right)\xi^j\epsilon^k t - L^j_{bc}\xi^b\epsilon^c\left(\frac{\partial x'_i}{\partial x_j}\right)t + O(t^2).\end{aligned}$$

Hence, if $$d\xi'^i = -L'^i_{jk}\xi'^j\epsilon'^k t + O(t^2),$$

$$L'^i_{jk}\left(\frac{\partial x'_j}{\partial x_b}\right)\left(\frac{\partial x'_k}{\partial x_c}\right)\xi^b\epsilon^c = \left[L^j_{bc}\left(\frac{\partial x'_i}{\partial x_j}\right) - \left(\frac{\partial^2 x'_i}{\partial x_b\,\partial x_c}\right)\right]\xi^b\epsilon^c.$$

The object at P which consists of the association of a set of numbers L^i_{jk} with each allowable coordinate system having the law of transformation

$$L^j_{bc}\frac{\partial x'_i}{\partial x_j} = L'^i_{jk}\frac{\partial x'_j}{\partial x_b}\frac{\partial x'_k}{\partial x_c} + \frac{\partial^2 x'_i}{\partial x_b\,\partial x_c} \tag{3}$$

is called an *affine connection*. It will be observed that in order to define an affine connection, the manifold must be of class *two* at least. We now suppose that an affine connection is given at each point of the manifold, the components being functions of class $(u-2)$.

Let Q be any other point of the manifold and let

$$x_i = x_i(t) \qquad (0 \leqslant t \leqslant 1)$$

be an arc C joining P to Q. The differential equation

$$\frac{d\xi^i}{dt} + L^i_{jk}\xi^j\frac{dx_k}{dt} = 0$$

has a unique solution ξ^i reducing to ξ^i_0 at $t = 0$. Let ξ^i_1 be the value of this solution at $t = 1$. We say that the vector ξ^i_1 at Q is the parallel displacement with respect to C of the vector ξ^i_0 at P. It is unaltered by a transformation of the coordinate system.

8·2. So far we have only considered contravariant vectors. Let us now consider the scalar ϕ at P of weight W. We define the infinitesimal parallel displacement of ϕ to P' as $\phi + d\phi$, where

$$d\phi = W\phi L^i_{ik}\epsilon^k t + O(t^2).$$

But it is necessary to show that this definition is independent of the coordinate system. For any transformation of coordinates we have

$$
\begin{aligned}
&\phi' + d\phi' \\
&= \left(\left|\frac{\partial x_i}{\partial x'_j}\right|^W\right)_{P'} (\phi + d\phi) \\
&= \left(\left|\frac{\partial x_i}{\partial x'_j}\right|^W\right)\left[1 + W\left(\frac{\partial^2 x_i}{\partial x'_j\,\partial x'_k}\right)\left(\frac{\partial x'_j}{\partial x_i}\right)\epsilon'^k t\right](\phi + d\phi) + O(t^2) \\
&= \left(\left|\frac{\partial x_i}{\partial x'_j}\right|^W\right)\left[1 + W\left(\frac{\partial^2 x_i}{\partial x'_j\,\partial x'_k}\right)\left(\frac{\partial x'_j}{\partial x_i}\right)\epsilon'^k t\right]\phi\,[1 + W L^i_{ik}\epsilon^k t] + O(t^2) \\
&= \phi'\left[1 + W\left(\frac{\partial^2 x_i}{\partial x'_j\,\partial x'_k}\frac{\partial x'_j}{\partial x_i} + L^i_{ij}\frac{\partial x_j}{\partial x'_k}\right)\epsilon'^k t\right] + O(t^2) \\
&= \phi' + W\phi' L'^i_{ik}\epsilon'^k t + O(t^2).
\end{aligned}
$$

This proves that the definition is independent of the coordinate system. We now define parallel displacement of a scalar along the arc C by means of the differential equation

$$\frac{d\phi}{dt} = W\phi L^i_{ik}\frac{dx_k}{dt}.$$

8·3. The definitions of the infinitesimal parallel displacement of a contravariant vector and of a scalar of weight W enable us to fix uniquely a parallelism between the tensors at P and at P'. If η_i is an absolute covariant vector, and ξ^i is an absolute contravariant vector, $\eta_i\xi^i$ is an absolute scalar. Hence the parallel displacement is given by

$$
\begin{aligned}
0 = d(\eta_i\xi^i) &= \eta_i d\xi^i + \xi^i d\eta_i + O(t^2) \\
&= (d\eta_i - L^k_{ij}\eta_k\epsilon^j t)\,\xi^i + O(t^2),
\end{aligned}
$$

and since this is to be true for all vectors ξ^i we must have

$$d\eta_i = L^k_{ij}\eta_k\epsilon^j t + O(t^2)$$

as the equation defining the infinitesimal parallel displacement of η_i. Next, let $T^{j_1\ldots j_q}_{i_1\ldots i_p}$ be any tensor of weight W at P. The infinitesimal parallel displacement $T^{j_1\ldots j_q}_{i_1\ldots i_p} + dT^{j_1\ldots j_q}_{i_1\ldots i_p}$ is found as follows. Let $\eta_{(1)i}, \ldots, \eta_{(q)i}$ be any q absolute covariant vectors and $\xi^i_{(1)}, \ldots, \xi^i_{(p)}$ be any p absolute contravariant vectors. Then

$$T^{j_1 \ldots j_q}_{i_1 \ldots i_p} \eta_{(1)j_1} \cdots \eta_{(q)j_q} \xi^{i_1}_{(1)} \cdots \xi^{i_p}_{(p)}$$

is a scalar of weight W. By the condition (ii) for parallelism we must have

$$0 = d(T^{j_1 \ldots j_q}_{i_1 \ldots i_p} \eta_{(1)j_1} \cdots \eta_{(q)j_q} \xi^{i_1}_{(1)} \cdots \xi^{i_p}_{(p)})$$
$$- W L^i_{ik} \epsilon^k t \, T^{j_1 \ldots j_q}_{i_1 \ldots i_p} \eta_{(1)j_1} \cdots \eta_{(q)j_q} \xi^{i_1}_{(1)} \cdots \xi^{i_p}_{(p)} + O(t^2).$$

But $$d\eta_{(r)i} = L^k_{ij} \eta_{(r)k} \epsilon^j t + O(t^2),$$

and $$d\xi^i_{(s)} = - L^i_{jk} \xi^j_{(s)} \epsilon^k t + O(t^2).$$

Hence

$$0 = \Big[dT^{j_1 \ldots j_q}_{i_1 \ldots i_p} + \sum_{r=1}^{q} T^{j_1 \ldots j_{r-1} k j_{r+1} \ldots j_q}_{i_1 \ldots \ldots \ldots \ldots \ldots \ldots i_p} L^{j_r}_{ka} \epsilon^a t$$
$$- \sum_{s=1}^{p} T^{j_1 \ldots \ldots \ldots \ldots \ldots \ldots j_q}_{i_1 \ldots i_{s-1} k i_{s+1} \ldots i_p} L^k_{i_s a} \epsilon^a t$$
$$- W L^i_{ia} \epsilon^a t \, T^{j_1 \ldots j_q}_{i_1 \ldots i_p} \Big] \eta_{(1)j_1} \cdots \eta_{(q)j_q} \xi^{i_1}_{(1)} \cdots \xi^{i_p}_{(p)} + O(t^2).$$

Since this is to hold whatever the vectors $\eta_{(1)i}, \ldots, \eta_{(q)i}, \xi^i_{(1)}, \ldots, \xi^i_{(p)}$ are, we obtain as the equation for parallel displacement of a tensor

$$dT^{j_1 \ldots j_q}_{i_1 \ldots i_p} + \sum_{r=1}^{q} T^{j_1 \ldots j_{r-1} k j_{r+1} \ldots j_q}_{i_1 \ldots \ldots \ldots \ldots \ldots \ldots i_p} L^{j_r}_{ka} \delta x^a - \sum_{s=1}^{p} T^{j_1 \ldots \ldots \ldots \ldots \ldots \ldots j_q}_{i_1 \ldots i_{s-1} k i_{s+1} \ldots i_p} L^k_{i_s a} \delta x^a$$
$$- W L^i_{ia} \delta x^a \, T^{j_1 \ldots j_q}_{i_1 \ldots i_p} = 0. \qquad (4)$$

Conversely, if the infinitesimal parallel displacement of a tensor is defined by this equation, the correspondence between the vectors at P and at P' is a parallelism. When the affine connection is defined throughout the manifold, we define the parallel displacement of a tensor along any arc C by means of a differential equation, as in the special case of a contravariant vector.

There are more general ways of defining parallel displacement, but we shall not have occasion to consider them.

9·1. Covariant differentiation. We suppose that an affine connection is given at all points of the manifold, the components in any coordinate system being of class $(u-2)$. Up to the present, we have considered tensors defined at a point, and parallel displacements. We have now to consider a field of tensors, that is, a set of tensors associated with the points of a region of the manifold, the components of the terms being functions of class v $(0<v<u)$ of the points of the manifold. For example, the metrical tensor g_{ij} belongs to a tensor field of class $(u-1)$.

Let us consider a tensor field $T^{j_1 \dots j_q}_{i_1 \dots i_p}$ of weight W. The law of transformation (2) holds at all points where the field is defined. If we differentiate this equation with respect to x'_k we obtain

$$\frac{\partial T'^{j_1 \dots j_q}_{i_1 \dots i_p}}{\partial x'_k} = \left|\frac{\partial x_i}{\partial x'_j}\right|^W \frac{\partial T^{b_1 \dots b_q}_{a_1 \dots a_p}}{\partial x_c} \frac{\partial x_{a_1}}{\partial x'_{i_1}} \cdots \frac{\partial x_{a_p}}{\partial x'_{i_p}} \frac{\partial x_c}{\partial x'_k} \frac{\partial x'_{j_1}}{\partial x_{b_1}} \cdots \frac{\partial x'_{j_q}}{\partial x_{b_q}}$$

$$+ W \left|\frac{\partial x_i}{\partial x'_j}\right|^W \frac{\partial^2 x_i}{\partial x'_j \partial x'_k} \frac{\partial x'_j}{\partial x_i} T^{b_1 \dots b_q}_{a_1 \dots a_p} \frac{\partial x_{a_1}}{\partial x'_{i_1}} \cdots \frac{\partial x_{a_p}}{\partial x'_{i_p}} \frac{\partial x'_{j_1}}{\partial x_{b_1}} \cdots \frac{\partial x'_{j_q}}{\partial x_{b_q}}$$

$$+ \left|\frac{\partial x_i}{\partial x'_j}\right|^W \sum_{s=1}^{p} T^{b_1 \dots\dots\dots\dots\dots\dots b_q}_{a_1 \dots a_{s-1} c\, a_{s+1} \dots a_p} \frac{\partial x_{a_1}}{\partial x'_{i_1}} \cdots \frac{\partial x_{a_{s-1}}}{\partial x'_{i_{s-1}}} \frac{\partial^2 x_c}{\partial x'_{i_s} \partial x'_k}$$

$$\times \frac{\partial x_{a_{s+1}}}{\partial x'_{i_{s+1}}} \cdots \frac{\partial x_{a_p}}{\partial x'_{i_p}} \frac{\partial x'_{j_1}}{\partial x_{b_1}} \cdots \frac{\partial x'_{j_q}}{\partial x_{b_q}}$$

$$+ \left|\frac{\partial x_i}{\partial x'_j}\right|^W \sum_{r=1}^{q} T^{b_1 \dots b_{r-1} c\, b_{r+1} \dots b_q}_{a_1 \dots\dots\dots\dots\dots\dots a_p} \frac{\partial x_{a_1}}{\partial x'_{i_1}} \cdots \frac{\partial x_{a_p}}{\partial x'_{i_p}} \frac{\partial x'_{j_1}}{\partial x_{b_1}} \cdots \frac{\partial x'_{j_{r-1}}}{\partial x_{b_{r-1}}}$$

$$\times \frac{\partial^2 x'_{j_r}}{\partial x_c \partial x_d} \frac{\partial x_d}{\partial x'_k} \frac{\partial x'_{j_{r+1}}}{\partial x_{b_{r+1}}} \cdots \frac{\partial x'_{j_q}}{\partial x_{b_q}}.$$

The derivative is not a tensor, since the equations of transformation contain partial derivatives of the second order. By equation (3), and the equation obtained from it by interchanging the dashed and undashed letters, we can eliminate the second derivatives and obtain an equation which involves the components of the affine connection in the two coordinate systems. On reducing the equation so obtained, we get

$$T'^{j_1\dots j_q}_{i_1\dots i_p,k} = \left|\frac{\partial x_i}{\partial x'_j}\right|^W T^{b_1\dots b_q}_{a_1\dots a_p,c}\frac{\partial x_{a_1}}{\partial x'_{i_1}}\cdots\frac{\partial x_{a_p}}{\partial x'_{i_p}}\frac{\partial x_c}{\partial x'_k}\frac{\partial x'_{j_1}}{\partial x_{b_1}}\cdots\frac{\partial x'_{j_q}}{\partial x_b},$$

where

$$T^{j_1\dots j_q}_{i_1\dots i_p,k} = \frac{\partial T_{i_1\dots i_p}}{\partial x_k} + \sum_{r=1}^{q} T^{j_1\dots j_{r-1}cj_{r+1}\dots j_q}_{i_1\dots\dots\dots\dots\dots\dots i_p} L^{j_r}_{ck}$$

$$-\sum_{s=1}^{p} T^{j_1\dots\dots\dots\dots\dots\dots j_q}_{i_1\dots i_{s-1}c\,i_{s+1}\dots i_p} L^c_{i_sk} - WT^{j_1\dots j_q}_{i_1\dots i_p} L^i_{ik}, \quad (5)$$

and $T'^{j_1\dots j_q}_{i_1\dots i_p,k}$ is similarly defined.

Thus, by means of the affine connection we can deduce from the partial derivative of a tensor another tensor, of the same weight but with one more covariant index. This we call the *covariant derivative* of the tensor. It will be noted that if P is the point $(\bar{x}_1, \dots, \bar{x}_n)$ and P' is $(\bar{x}_1 + \delta x^1, \dots, \bar{x}_n + \delta x^n)$, the infinitesimal parallel displacement to P' of the tensor $T^{j_1\dots j_q}_{i_1\dots i_p}$ at P belonging to a tensor field is

$$T^{j_1\dots j_q}_{i_1\dots i_p}(P') - T^{j_1\dots j_q}_{i_1\dots i_p,k}\,\delta x^k.$$

We prove immediately that

$$[T^{j_1\dots j_q}_{i_1\dots i_p} + S^{j_1\dots j_q}_{i_1\dots i_p}]_{,k} = T^{j_1\dots j_q}_{i_1\dots i_p,k} + S^{j_1\dots j_q}_{i_1\dots i_p,k},$$

and

$$[T^{j_1\dots j_q}_{i_1\dots i_p} S^{m_1\dots m_s}_{l_1\dots\dots l_r}]_{,k} = T^{j_1\dots j_q}_{i_1\dots i_p,k} S^{m_1\dots m_s}_{l_1\dots\dots l_r} + T^{j_1\dots j_q}_{i_1\dots i_p} S^{m_1\dots m_s}_{l_1\dots\dots l_r,k}.$$

9·2. Provided that the class number v is sufficiently high, we can repeat the process of forming the covariant derivative of a tensor in a tensor field. But in general the second covariant derivative is not symmetrical in the indices which denote differentiation. This can be verified by direct calculation. We omit the details of the calculation, which is quite straightforward, and merely state the formula:

$$T^{j_1\ldots j_q}_{i_1\ldots i_p,s,t} - T^{j_1\ldots j_q}_{i_1\ldots i_p,t,s} = \sum_{r=1}^{q} T^{j_1\ldots j_{r-1}\,a\,j_{r+1}\ldots j_q}_{i_1\ldots\ldots\ldots\ldots\ldots i_p} B^{j_r}_{ast}$$

$$- \sum_{r=1}^{p} T^{j_1\ldots\ldots\ldots\ldots\ldots j_q}_{i_1\ldots i_{r-1}\,a\,i_{r+1}\ldots i_p} B^{a}_{i_r st} - 2T^{j_1\ldots j_q}_{i_1\ldots i_p,k}\Omega^k_{st} - WT^{j_1\ldots j_q}_{i_1\ldots i_p} B^a_{ast},$$

where $$2\Omega^k_{st} = L^k_{st} - L^k_{ts}$$

and $$B^i_{rst} = \frac{\partial L^i_{rs}}{\partial x_t} - \frac{\partial L^i_{rt}}{\partial x_s} + L^a_{rs}L^i_{at} - L^a_{rt}L^i_{as}. \qquad (6)$$

From the equation (3) we prove that Ω^k_{st} is an absolute tensor. We subtract from (3) the equation formed by permuting b and c. The resulting equation gives the law of transformation of Ω^k_{bc}, which is that of a tensor.

Again, B^i_{rst} is also an absolute tensor. To prove this, take $T^{j_1\ldots j_q}_{i_1\ldots i_p}$ to be an absolute contravariant vector ξ^i. Since

$$\xi^i{}_{,s,t} - \xi^i{}_{,t,s} + 2\xi^i{}_{,k}\Omega^k_{st}$$

is a tensor, so is $\xi^a B^i_{ast}$. We therefore have the transformation law

$$\xi^b \frac{\partial x'_a}{\partial x_b} B'^i_{ast} = \xi'^a B'^i_{ast} = \xi^a B^l_{amn} \frac{\partial x'_i}{\partial x_l} \frac{\partial x_m}{\partial x'_s} \frac{\partial x_n}{\partial x'_t}.$$

Since this holds for all vectors ξ^i, we have

$$B'^i_{rst} = B^l_{bmn} \frac{\partial x'_i}{\partial x_l} \frac{\partial x_b}{\partial x'_r} \frac{\partial x_m}{\partial x'_s} \frac{\partial x_n}{\partial x'_t},$$

that is, B^i_{rst} is a tensor of weight zero. We call it the *curvature tensor*.

9·3. Direct calculation shows that the covariant derivatives of the numerical tensors are zero. Consider, for example, $\epsilon_{i_1\ldots i_n}$. We have

$$\epsilon_{i_1\ldots i_n,k} = -\sum_{s=1}^{n} \epsilon_{i_1\ldots i_{s-1}\,a\,i_{s+1}\ldots i_n} L^a_{i_s k} + \epsilon_{i_1\ldots i_n} L^i_{ik} = 0.$$

The results for $\epsilon^{i_1\ldots i_n}$ and $\delta^{j_1\ldots j_p}_{i_1\ldots i_p}$ are proved similarly.

10·1. Riemannian geometry. Riemannian geometry is defined by the properties:

(i) there exists a fundamental metrical tensor field;

(ii) the affine connection is symmetric in the two lower indices;

(iii) the length $\{g_{ij}\xi^i\xi^j\}^{\frac{1}{2}}$

of any absolute contravariant vector is unaltered by infinitestimal parallel displacement.

We have defined Ω^k_{st} as the skew-symmetric part of the affine connection. We denote the symmetric part by Γ^k_{st}, so that

$$L^k_{st} = \Gamma^k_{st} + \Omega^k_{st}.$$

In Riemannian geometry we have

$$L^k_{st} = \Gamma^k_{st}.$$

From the invariance of the length of ξ^i we have

$$g_{ij}(x+\delta x)(\xi^i + d\xi^i)(\xi^j + d\xi^j) = g_{ij}\xi^i\xi^j,$$

where $g_{ij}(x+\delta x)$ denotes the value of g_{ij} at $(x_1+\delta x^1, \ldots, x_n+\delta x^n)$, and

$$d\xi^i = -L^i_{jk}\xi^j\delta x^k.$$

Hence $$\left[\frac{\partial g_{ij}}{\partial x_k} - g_{aj}\Gamma^a_{ik} - g_{ia}\Gamma^a_{jk}\right]\xi^i\xi^j\delta x^k = 0,$$

that is, $$g_{ij,k}\xi^i\xi^j\delta x^k = 0.$$

Since this holds for all vectors ξ^i, and for all displacements,

$$g_{ij,k} = \frac{\partial g_{ij}}{\partial x_k} - g_{aj}\Gamma^a_{ik} - g_{ia}\Gamma^a_{jk} = 0.$$

Solving this equation, we obtain

$$\Gamma^i_{jk} = \tfrac{1}{2}g^{ia}\left[\frac{\partial g_{aj}}{\partial x_k} + \frac{\partial g_{ak}}{\partial x_j} - \frac{\partial g_{jk}}{\partial x_a}\right].$$

The components of affine connection in Riemannian geometry are called *Christoffel symbols*.

10·2. From the equations

$$g^{ij}g_{jk} = \delta^i_k, \quad g_{ij,t} = 0, \quad \delta^i_{k,t} = 0,$$

we have

$$g^{ij}{}_{,t}\,g_{jk} = 0,$$

hence

$$g^{ij}{}_{,t}\,g_{jk}\,g^{ks} = g^{ij}{}_{,t}\,\delta^s_j$$
$$= g^{is}{}_{,t} = 0.$$

Again

$$g_{,k} = \frac{\partial g}{\partial x_k} - 2g\Gamma^i_{ik}$$
$$= g\left[\frac{\partial g_{ij}}{\partial x_k}g^{ij} - 2\Gamma^i_{ik}\right]$$
$$= g\,[g_{aj}\Gamma^a_{ik}g^{ij} + g_{ia}\Gamma^a_{jk}g^{ij} - 2\Gamma^i_{ik}]$$
$$= g\,[\Gamma^i_{ik} + \Gamma^j_{jk} - 2\Gamma^i_{ik}]$$
$$= 0,$$

and, similarly,

$$(\sqrt{g})_{,k} = 0.$$

From this last equation, which we write as

$$\frac{\partial\sqrt{g}}{\partial x_k} - \sqrt{g}\,\Gamma^i_{ik} = 0,$$

we deduce that in Riemannian geometry

$$\Gamma^i_{ik} = \frac{\partial}{\partial x_k}\{\log\sqrt{g}\}.$$

Hence

$$B^i_{ist} = \frac{\partial\Gamma^i_{is}}{\partial x_t} - \frac{\partial\Gamma^i_{it}}{\partial x_s} = 0,$$

and, since we have, by definition,

$$\Omega^k_{st} = 0,$$

the last two terms in the formula for the interchange of the order of covariant differentiation are absent.

10·3. There are certain identities satisfied by the curvature tensor which we must now prove.

(i) For all affine connections

$$B^a_{ijk} = -B^a_{ikj}.$$

(ii) For all symmetric affine connections the formula (6) leads at once to

$$B^a_{ijk} + B^a_{jki} + B^a_{kij} = 0.$$

(iii) In Riemannian space it is convenient to introduce a covariant tensor of rank four, which we call the *Riemann-Christoffel tensor*:

$$R_{ijkl} = g_{ia} B^a_{jkl}.$$

Clearly, $$R_{ijkl} = -R_{ijlk},$$

and $$R_{ijkl} + R_{iklj} + R_{iljk} = 0.$$

Since $$g_{ij,s,t} = 0 = g_{ij,t,s},$$

we have $$g_{aj} B^a_{ist} + g_{ia} B^a_{jst} = 0,$$

and therefore $$R_{jist} + R_{ijst} = 0.$$

Now add the two equations

$$R_{ijkl} + R_{iklj} + R_{iljk} = 0,$$

$$R_{jikl} + R_{jkli} + R_{jlik} = 0.$$

We get $$(R_{iklj} + R_{jkli}) + (R_{iljk} + R_{jlik}) = 0,$$

and, by the identities already obtained, we have

$$(R_{kilj} + R_{kjli}) + (R_{lijk} + R_{ljik}) = 0,$$

and $$2R_{kjli} + R_{klij} + 2R_{lijk} + R_{lkij} = 0;$$

and therefore $$R_{kjli} = R_{likj}.$$

10·4. In § 3·1 we defined the length of an arc in Riemannian space. Consider an arc C, and express the coordinates of points on it as functions of the distance s measured from a fixed point of C. At each point of C there is defined a contravariant vector ξ^i, called the *tangent vector*, whose components are given by

$$\xi^i = \frac{dx_i}{ds}.$$

The arc C is called a *geodesic* if the tangent vector satisfies the equation

$$\frac{d\xi^i}{ds}+\Gamma^i_{jk}\xi^j\xi^k = 0.$$

It can be proved that through any point P of the space there passes a unique geodesic whose tangent at P is any assigned contravariant vector, and that there exists a neighbourhood of P with the property that each point Q of this neighbourhood can be joined to P by a unique geodesic lying entirely in the neighbourhood. The length of this geodesic is called the *geodesic distance* between P and Q, and the locus of points at a constant geodesic distance from P is called a *geodesic sphere.*

11. **Geodesic coordinates.** The affine connection is not a tensor. We have seen that if the components of a tensor vanish in one system of coordinates they vanish in every system of coordinates. This is not the case for the affine connection.

Given the components of affine connection in the coordinate system $(x_1, \ldots, x_n)$, equation (3) shows that in order to find a coordinate system $(x'_1, \ldots, x'_n)$ in which $L'^i_{jk} = 0$ we have to solve the equations

$$\frac{\partial^2 x'_i}{\partial x_j\, \partial x_k} = L^a_{jk}\frac{\partial x'_i}{\partial x_a}.$$

The necessary and sufficient condition that there should exist a coordinate system in which the components of affine connection vanish is that these equations should be completely integrable. The conditions for this are found to be

$$L^a_{jk} = L^a_{kj} = \Gamma^a_{jk},$$

and

$$B^a_{ijk} = 0.$$

The first condition is satisfied in Riemannian geometry, but in general the second is not. When it is satisfied the Riemannian geometry is said to be flat, and it can be shown that the geometry is equivalent to ordinary Euclidean geometry.

The local geometry of a Riemannian manifold is not, however, flat, so we cannot find a coordinate system in which the components of affine connection vanish at the points of a neighbourhood. But, let us consider the transformation of coordinates

$$x'_i = x_i - \bar{x}_i + \tfrac{1}{2} \sum_{jk} (\Gamma^i_{jk})_{\bar{x}} (x_j - \bar{x}_j)(x_k - \bar{x}_k).$$

Equation (3) shows that in the coordinate system $(x'_1, \ldots, x'_n)$ the components of affine connection vanish at the origin of coordinates. Now make a further linear transformation

$$x''_i = \sum_j a_{ij} x'_j$$

of the coordinates, so that

$$\sum_{i,j=1}^{n} g'_{ij}(0)\, x'_i x'_j = \sum_{i=1}^{n} (x''_i)^2.$$

In the coordinate system $(x''_1, \ldots, x''_n)$ the components of affine connection vanish at the origin, and at the origin we have

$$g''_{ii} = 1, \quad g''_{ij} = 0 \ (i \neq j), \quad \frac{\partial g''_{ij}}{\partial x''_k} = 0.$$

A coordinate system with these two properties is called a *geodesic coordinate system* at the point which is the origin of the coordinates. The use of geodesic coordinate systems is convenient since, when we are performing calculations, the equations which we obtain are simpler than in the case when general coordinate systems are used. The following facts, which hold when we deal with geodesic coordinates, will be of considerable use in later chapters:

(i) the first covariant derivative of a tensor is equal to the ordinary derivative at the origin of geodesic coordinates;

(ii) at the origin of geodesic coordinates

$$R_{ijkl} = \frac{1}{2}\left(\frac{\partial^2 g_{ik}}{\partial x_j \partial x_l} - \frac{\partial^2 g_{il}}{\partial x_j \partial x_k} - \frac{\partial^2 g_{jk}}{\partial x_i \partial x_l} + \frac{\partial^2 g_{jl}}{\partial x_i \partial x_k}\right).$$

TOPOLOGY

In the following paragraphs we give a brief account of the more important topological properties of Riemannian manifolds, which will frequently be used in later chapters. The proofs of the theorems which will be stated can be found in standard treatises on topology, and it is unnecessary to repeat them here. The reader is therefore referred to the standard works, such as those of Lefschetz [5], Seifert and Threlfall [8], or Alexandroff and Hopf [1]. In the account which follows, explicit reference to these works will not be made; but in the case of certain results not proved in these books, references will be given to the original papers in which the theorems are proved.

12·1. **Polyhedral complexes.** In the Euclidean space $\mathscr{E}_r$ $(x_1, \ldots, x_r)$, consider $p+1$ independent points $P_0, \ldots, P_p$. If P_k has coordinates $(x_1^k, \ldots, x_r^k)$, the matrix

$$\begin{pmatrix} x_1^0 & \ldots & x_r^0 & 1 \\ x_1^1 & \ldots & x_r^1 & 1 \\ \vdots & & & \\ x_1^p & \ldots & x_r^p & 1 \end{pmatrix}$$

is of rank $p+1$. The set of points given by

$$x_i = \lambda_k x_i^k \qquad (i = 1, \ldots, r),$$

where

$$\lambda_k > 0 \qquad (k = 0, \ldots, p)$$

and

$$\lambda_0 + \ldots + \lambda_p = 1,$$

forms a *rectilinear p-simplex*, or a *rectilinear simplex of p dimensions*, which we denote by $(P_0 \ldots P_p)$. If we replace the inequalities

$$\lambda_k > 0$$

by

$$\lambda_k \geqslant 0,$$

the set of points obtained is called the *closure* of $(P_0 \ldots P_p)$, and the difference is the *boundary* of the simplex. The boundary therefore consists of $\binom{p+1}{k+1}$ simplexes of dimension k, for $k = 0, \ldots, p-1$.

We now consider a finite set of rectilinear simplexes in $\mathscr{E}_r$, of dimensions $0, \ldots, n$, having the properties:

(i) no two simplexes of the set have a point in common;

(ii) every simplex lying on the boundary of a simplex of the set belongs to the set.

Such a set of simplexes forms a *polyhedral complex* of n dimensions. Thus the simplexes forming the faces, edges, and vertices of a tetrahedron form a polyhedral complex of two dimensions.

If $(P_0 \ldots P_p)$ is any p-simplex of the set, it can equally well be denoted by $(P_{i_0} \ldots P_{i_p})$, where $(i_0, \ldots, i_p)$ is any derangement of the numbers $(0, \ldots, p)$. These derangements fall into two classes, those which are obtained from $(0, \ldots, p)$ by an even permutation forming one class, and those obtained by an odd permutation forming the other. Any derangement is obtained from another of the same class by an even permutation, and from a derangement of the other class by an odd permutation. We now define an *oriented* simplex from $(P_0 \ldots P_p)$ by associating with it one of the classes of derangements of the suffixes, and we denote the oriented simplex by $P_{i_0} \ldots P_{i_p}$ (without the brackets), where $(i_0, \ldots, i_p)$ is a derangement of the associated class. The oriented simplex obtained by associating with $(P_0 \ldots P_p)$ the other class of derangements is denoted by $P_{j_0} \ldots P_{j_p}$, where $(j_0, \ldots, j_p)$ is an appropriate derangement of $(0, \ldots, p)$, but we also denote it by $-P_{i_0} \ldots P_{i_p}$.

If we orient (in an arbitrary manner) all the simplexes of a complex, we obtain an *oriented complex*. We introduce the ideas with which we shall deal by considering certain properties of an oriented polyhedral complex.

12·2. We denote the oriented p-simplexes of the polyhedral complex K by E_p^i $(i = 1, \ldots, \alpha_p)$, for $p = 0, \ldots, n$.

We take the oriented p-simplexes E_p^i $(i = 1, \ldots, \alpha_p)$ as the free generators of an additive group. The general member of the group can be written as

$$C_p = a_i E^i_p,$$

where a_i is any integer (and we make use of the summation convention). C_p is called a *p-chain*.

Consider the p-chain E^i_p, equal to the oriented simplex $P_{i_0} \ldots P_{i_p}$. We define the boundary of this to be the $(p-1)$-chain

$$F(E^i_p) = C_{p-1} = \sum_{j=0}^{p} (-1)^j P_{i_0} \ldots P_{i_{j-1}} P_{i_{j+1}} \ldots P_{i_p},$$

and it can be verified at once that this boundary does not depend on the particular derangement $(i_0 \ldots i_p)$ chosen within the class of derangements which orient the p-simplex. The boundary of $-P_{i_0} \ldots P_{i_p}$ is $-C_{p-1}$. The boundary can be written as

$$F(E^i_p) = {}_{(p)}\eta^i_j E^j_{p-1},$$

where

$${}_{(p)}\eta^i_j = -1, 0, \text{ or } 1,$$

and the summation with respect to j is from $j = 1$ to $j = \alpha_{p-1}$. The number ${}_{(p)}\eta^i_j$ is thus defined for all i, j $(1 \leqslant i \leqslant \alpha_p;$ $1 \leqslant j \leqslant \alpha_{p-1})$, and for all values of p $(1 \leqslant p \leqslant n)$. We write the relation between E^i_p and its boundary chain as

$$E^i_p \to {}_{(p)}\eta^i_j E^j_{p-1}.$$

These relations, for $i = 1, \ldots, \alpha_p$, form the *incidence relations* for the p-simplexes of K, and the matrix

$${}_{(p)}\boldsymbol{\eta} = ({}_{(p)}\eta^i_j)$$

is called the pth *incidence matrix*. If

$$C_p = a_i E^i_p$$

is any p-chain, we define its boundary as

$$C_{p-1} = a_i\, {}_{(p)}\eta^i_j E^j_{p-1}.$$

Now

$$P_{i_0}\dots P_{i_p} \to \sum_{j=0}^{p} (-1)^j P_{i_0}\dots P_{i_{j-1}} P_{i_{j+1}}\dots P_{i_p}$$

$$\to \sum_{j=0}^{p} (-1)^j \left[\sum_{k=0}^{j-1} (-1)^k P_{i_0}\dots P_{i_{k-1}} P_{i_{k+1}}\dots P_{i_{j-1}} P_{i_{j+1}}\dots P_{i_p} \right.$$

$$\left. + \sum_{k=j+1}^{p} (-1)^{k-1} P_{i_0}\dots P_{i_{j-1}} P_{i_{j+1}}\dots P_{i_{k-1}} P_{i_{k+1}}\dots P_{i_p} \right]$$

$$= 0.$$

Hence $${}_{(p)}\eta^i_j E^j_p \to {}_{(p)}\eta^i_j\, {}_{(p-1)}\eta^j_k E^k_{p-2} = 0,$$

that is, $${}_{(p)}\eta \cdot {}_{(p-1)}\eta = 0, \qquad (7)$$

and, in general, if C_p is any chain,

$$F(C_p) \to 0.$$

12·3. A p-chain whose boundary is zero is called a *p-cycle*. The result which we have just established tells us that the boundary of any $(p+1)$-chain is a p-cycle, or else zero. But the converse result, that every p-cycle is the boundary of a $(p+1)$-chain, is not generally true. We therefore distinguish the cycles which are boundaries by calling them *bounding cycles*. If C_p is a bounding cycle, we say that it is *homologous to zero*, and write

$$C_p \sim 0.$$

If C_p and C'_p are two p-chains such that

$$C_p - C'_p \sim 0,$$

we write $$C_p \sim C'_p.$$

In order to determine the p-chains of K which are p-cycles, and to find which of the cycles are bounding cycles, we make a change of base for the groups of $(p+1)$-chains, p-chains, and $(p-1)$-chains. By using (7), and an elementary result in matrix algebra, we can show that new bases

$$C^i_{p+1}\,(i = 1, \dots, \alpha_{p+1}),\quad C^j_p\,(j = 1, \dots, \alpha_p),\quad C^k_{p-1}\,(k = 1, \dots, \alpha_{p-1})$$

can be found so that the incidence relations can be written in the following form (in which the summation convention does *not* apply):

$$\left.\begin{array}{ll} C^i_{p+1} \to d_i C^i_p & (i = 1, \ldots, \rho_{p+1}), \\ C^i_{p+1} \to 0 & (i = \rho_{p+1}+1, \ldots, \alpha_{p+1}); \\ C^j_p \to 0 & (j = 1, \ldots, \rho_{p+1}), \\ C^{\rho_{p+1}+j}_p \to e_j C^j_{p-1} & (j = 1, \ldots, \rho_p), \\ C^j_p \to 0 & (j = \rho_p + \rho_{p+1} + 1, \ldots, \alpha_p). \end{array}\right\} \quad (8)$$

In these equations, ρ_p is the rank of ${}_{(p)}\eta$ and $e_1 \geqslant e_2 \geqslant \ldots \geqslant e_{\rho_p}$ are its invariant factors; while $\rho_{p+1}, d_1, \ldots, d_{\rho_{p+1}}$ are similarly related to ${}_{(p+1)}\eta$. If Γ_p is any p-cycle of K,

$$\Gamma_p = \sum_{j=1}^{\rho_{p+1}} a_j C^j_p + \sum_{j=\rho_p+\rho_{p+1}+1}^{\alpha_p} b_j C^j_p,$$

where a_j, b_j are integers. If Γ_p is the boundary of a chain $\sum_1^{\alpha_{p+1}} c_i C^i_{p+1}$, then

$$\sum_{i=1}^{\rho_{p+1}} c_i d_i C^i_p = \sum_{j=1}^{\rho_{p+1}} a_j C^j_p + \sum_{j=\rho_p+\rho_{p+1}+1}^{\alpha_p} b_j C^j_p$$

and hence $\quad b_j = 0 \quad (j = \rho_p + \rho_{p+1} + 1, \ldots, \alpha_p),$

and $\quad a_j \equiv 0 \pmod{d_j} \quad (j = 1, \ldots, \rho_{p+1}).$

Conversely, if $\quad b_j = 0 \quad (j = \rho_p + \rho_{p+1} + 1, \ldots, \alpha_p)$

and $\quad a_j = c_j d_j \quad (j = 1, \ldots, \rho_{p+1}),$

Γ_p is the boundary of $\sum_{i=1}^{\rho_{p+1}} c_i C^i_{p+1}$.

The p-cycles of K form an additive group G_1, which is a sub-group of the additive group of p-chains. The group G_1 has a sub-group G_2 consisting of the boundary p-cycles, and the result which we have established shows that the factor group $H = G_1/G_2$ is generated by $R_p = \alpha_p - \rho_p - \rho_{p+1}$ free generators, corresponding to C^j_p $(j = \rho_p + \rho_{p+1} + 1, \ldots, \alpha_p)$, and θ_p generators of finite order corresponding to the cycles C^j_p $(j = 1, \ldots, \theta_p)$

where $d_1, \ldots, d_{\theta_p}$ are the invariant factors of ${}_{(p+1)}\eta$ which are greater than unity.

The group H is the direct sum of an infinite Abelian group H_1 having R_p free generators, and a finite Abelian group H_2 generated by θ_p generators of orders $d_1, \ldots, d_{\theta_p}$. The group H_1 is called the *Betti group* of K, and R_p is the pth *Betti number* of K. The group H_2 is called the pth *torsion* group of K, and $d_1, \ldots, d_{\theta_p}$ are called the pth *torsion coefficients* of K.

Any cycle Γ_p which can be written as

$$\Gamma_p = \sum_1^{\rho_{p+1}} a_i C_p^i$$

is either a bounding cycle, or else there exists an integer $\lambda > 1$ such that

$$\lambda \Gamma_p \sim 0.$$

In the latter case we say that Γ_p is a *divisor of zero*, and we say that Γ_p is *homologous to zero with division*, writing

$$\Gamma_p \approx 0.$$

If Γ_p and Γ'_p are two p-chains such that

$$\Gamma_p - \Gamma'_p \approx 0,$$

we say that they are homologous to each other with division, and write

$$\Gamma_p \approx \Gamma'_p.$$

12·4. Before leaving polyhedral complexes, we describe an operation on a complex K, known as regular sub-division. This consists in breaking up the simplexes of K into smaller simplexes, according to definite rules.

We define the process by induction on the dimension of the simplexes of K. In a 1-simplex $(P_0 P_1)$ take a point Q, and then replace $(P_0 P_1)$ by the 1-simplexes $(P_0 Q)$, $(Q P_1)$ and the 0-simplex Q. This gives a regular sub-division of $(P_0 P_1)$, and we can therefore form the regular sub-divisions of all the 1-simplexes of K. We now suppose that the process of regularly sub-dividing simplexes of dimension less than p has been defined, and that the q-simplexes of K have been so sub-divided, for $q = 1, \ldots, p-1$. Consider a p-simplex E_p^i, and in it take a point P_p^i. This point P_p^i can be joined to any simplex of the sub-divided boundary of E_p^i by a rectilinear simplex, and the sum of the simplexes so obtained, together with the 0-simplex P_p^i, exactly covers E_p^i. This gives the required sub-division of E_p^i. If we sub-divide each p-simplex of K in this way, we carry our process to the stage at which each simplex of dimension less than $p+1$ has been sub-divided. By continuing the induction we arrive at a stage at which all the simplexes of K are sub-divided. The new set of rectilinear simplexes forms a complex K' which is a *regular sub-division* of K. Each simplex of K is clearly covered by a set of simplexes of K'.

The simplexes of K' can be oriented in any way we please. It is sometimes convenient, however, to impose a condition on the method of orienting the new simplexes. Consider any p-simplex $(Q_0 \ldots Q_p)$ of K'. This may lie on a p-simplex $(P_0 \ldots P_p)$ of K. In this case, one $(p-1)$-simplex on its boundary lies on a $(p-1)$-simplex of the boundary of $(P_0 \ldots P_p)$. Let these $(p-1)$-simplexes be $(Q_{i_1} \ldots Q_{i_p})$, $(P_{j_1} \ldots P_{j_p})$. Then Q_{i_0} is the new vertex introduced into $(P_0 \ldots P_p)$ in forming its regular sub-division. We fix the orientation of $(Q_0 \ldots Q_p)$ inductively as follows.

(i) If $p = 1$, and the orientation of $(P_0 P_1)$ is given by $\xi P_0 P_1$ $(\xi = \pm 1)$, the orientation of the new 1-simplexes are given by $\xi P_0 Q$, $\xi Q P_1$.

(ii) We now suppose that the orientations of the $(p-1)$-simplexes of K' which lie on $(p-1)$-simplexes of K have been fixed, and that the oriented simplexes formed from $(Q_{i_1} \ldots Q_{i_p})$, $(P_{j_1} \ldots P_{j_p})$ are $\xi Q_{i_1} \ldots Q_{i_p}$, $\eta P_{j_1} \ldots P_{j_p}$ $(\xi, \eta = \pm 1)$. Then, if the oriented p-simplex of K is $\zeta P_{j_0} \ldots P_{j_p}$, the oriented p-simplex of K' is $\xi\eta\zeta Q_{i_0} \ldots Q_{i_p}$.

We can describe this process of orienting $(Q_0 \ldots Q_p)$ as follows. We suppose the vertices are arranged so that

$\xi = \eta = +1$. $(Q_0 \dots Q_p)$ and $(P_0 \dots P_p)$ lie in a p-space of the Euclidean space $\mathscr{E}_r$. We may suppose this space to be given by

$$x_{p+1} = 0, \dots, x_n = 0.$$

If P_k has coordinates $(x_1^k, \dots, x_n^k)$ and Q_k has coordinates $(y_1^k, \dots, y_n^k)$, condition (ii) is equivalent to saying that the determinants

$$\begin{vmatrix} x_1^{j_0} & \dots & x_p^{j_0} & 1 \\ \vdots & & & \\ x_1^{j_p} & \dots & x_p^{j_p} & 1 \end{vmatrix} \quad \text{and} \quad \begin{vmatrix} y_1^{i_0} & \dots & y_p^{i_0} & 1 \\ \vdots & & & \\ y_1^{i_p} & \dots & y_p^{i_p} & 1 \end{vmatrix}$$

have the same sign.

A p-simplex of K' may, on the other hand, lie on a simplex of K of dimension greater than p. In this case, no restriction on the orientations of the p-simplex of K' is imposed at present.

We shall adhere to this convention for orienting the simplexes of a regular sub-division. Its main advantage is that it simplifies the statement of some theorems.

12·5. We can now carry through the investigations made in §§ 12·2, 12·3, for the complex K'. By purely algebraic reasoning certain relations between the properties of K and the properties of K' can be obtained. The two following are of importance in the sequel:

(i) If the incidence relation

$$a_i E_{p+1}^i \rightarrow b_j E_p^j$$

holds in K, and in it we replace E_{p+1}^i by the chain-sum (with coefficients $+1$) of the $(p+1)$-simplexes of K' which lie on E_{p+1}^i, and if we similarly replace E_p^j by the chain-sum of the p-simplexes of K' on E_p^j, the incidence relation holds in K'.

(ii) The pth Betti number of K' is equal to the pth Betti number of K, and the pth torsion coefficients of K' are equal to the pth torsion coefficients of K. Thus the pth Betti group and torsion group of K' are isomorphic with the pth Betti group and torsion group of K.

13·1. Complexes of class v. The algebraic theory of complexes described in the preceding paragraphs does not depend in any essential respect on the fact that a p-simplex is a rectilinear simplex lying in a Euclidean space of p dimensions. Indeed, we could equally well apply it to any set of objects in (1-1) correspondence with the simplexes of a complex. In this paragraph we consider a generalisation which is required in order to apply the theory to manifolds.

Consider a rectilinear p-simplex $(P_0 \ldots P_p)$, and let $(u_1, \ldots, u_p)$ be a set of coordinates (e.g. cartesian coordinates) in the p-space in which it lies, valid in a region containing the simplex. Let $f_i(u_1, \ldots, u_p)$ $(i = 1, \ldots, N)$ be functions of $(u_1, \ldots, u_p)$ of class v in a region containing the simplex. Then, in the Euclidean space $(x_1, \ldots, x_N)$, consider the locus defined by the equations

$$x_i = f_i(u_1, \ldots, u_p) \qquad (i = 1, \ldots, N),$$

for all values of the parameters $(u_1, \ldots, u_p)$ in the simplex. We call this locus a *p-simplex of class v*. The simplex is said to be *non-singular* if

(i) the correspondence between its points and the points of $(P_0 \ldots P_p)$ is (1-1) without exception; and

(ii) when $v > 0$, the Jacobian matrix $\left(\frac{\partial f_i}{\partial u_j}\right)$ is of rank p at all points of $(P_0 \ldots P_p)$.

Otherwise the simplex is said to be *singular*.

A complex of class v is a finite set of simplexes of class v, such that the simplexes on the boundary of any simplex of the set belongs to the set. The complex is non-singular if its simplexes are all non-singular, and if no two simplexes have a point in common. It is clear that the notions of orientation, chains, and so on, can be carried over directly to complexes of class v, and the results of §§ 12·2–12·5 can be applied.

13·2. When the class v of a simplex is greater than zero, it is often convenient to define its orientation by means of parameters used to represent it. An oriented simplex E_p, of class

v, is defined as the image of an oriented rectilinear simplex, say, $P_0 \dots P_p$. Let $(u_1, \dots, u_p)$ be cartesian coordinates in the linear space containing $P_0 \dots P_p$, and let the coordinates of P_k be $(u_1^k, \dots, u_p^k)$. If the determinant

$$\begin{vmatrix} 1 & u_1^0 & \dots & u_p^0 \\ \vdots & & & \\ 1 & u_1^p & \dots & u_p^p \end{vmatrix}$$

is positive, we say that $(u_1, \dots, u_p)$ are parameters in E_p *concordant* with the orientation of E_p. If the determinant is negative, the parameters $(u_1, \dots, u_p)$ are concordant with the orientation of $-E_p$, and $(-u_1, u_2, \dots, u_p)$ are parameters concordant with the orientation of E_p.

Now consider any other set of parameters in E_p, given by

$$\bar{u}_i = \bar{u}_i(u_1, \dots, u_p) \qquad (i = 1, \dots, p),$$

where the functions $\bar{u}_1, \dots, \bar{u}_p$ do not take the same set of values at two different points of $(P_0 \dots P_p)$, and the Jacobian $\left|\dfrac{\partial \bar{u}_i}{\partial u_j}\right|$ is different from zero at all points of the simplex. We say that the parameters $(\bar{u}_1, \dots, \bar{u}_p)$ are concordant with the orientation of E_p if either (i) the parameters $(u_1, \dots, u_p)$ are concordant with the orientation of E_p and the Jacobian is positive, or (ii) the parameters $(u_1, \dots, u_p)$ are not concordant with the orientation and the Jacobian is negative. It is easily verified that this method of orienting the simplex by means of a set of concordant parameters is completely equivalent to the method of orienting it by an arrangement of the vertices. While the latter method is more convenient for the purely combinatorial theory of complexes, the use of parameters has certain advantages in the applications with which we shall deal in later chapters.

13·3. We now consider a non-singular complex K of class v. A complex $\bar{K}$ of class $\bar{v}$ is said to lie on K if the point-set formed by the points of its simplexes is contained in the point-set formed by the points of the simplexes of K. A chain $\bar{C}_p$ is said to lie on K if it is a chain of a complex lying on K. The fundamental theorem concerning chains and complexes which lie on a given complex K, of class v, is known as the *Deformation Theorem*. This theorem can be stated as follows. If C'_p is any chain of a complex K' of class v' lying on K, there exists a complex K'', obtained from K' by repeated regular sub-division, and a complex $\bar{K}$ on K, having K'' as a sub-complex, with the properties†:

(i) The class of $\bar{K}$ is equal to the smaller of v, v';

(ii) if $$C'_p \to C'_{p-1} \qquad (\text{in } K')$$
and if C''_p, C''_{p-1} are the chains of K'' obtained from C'_p, C'_{p-1} by the sub-division, there exist chains D_{p+1}, D_p of $\bar{K}$, such that
$$D_{p+1} \to C''_p - C_p + D_p \qquad (\text{in } \bar{K}),$$
and
$$D_p \to C_{p-1} - C''_{p-1} \qquad (\text{in } \bar{K}),$$
where C_p, C_{p-1} are chains of K, and
$$C_p \to C_{p-1} \qquad (\text{in } K);$$

(iii) if $C'_{p-1} = 0$, then $D_p = 0$, $C_{p-1} = 0$;

(iv) if C'_{p-1} is a cycle of K, then $D_p = 0$ and $C_{p-1} = C'_{p-1}$. C'_p is said to be *deformed* into C_p over D_{p+1} and C'_{p-1} *deformed* into C_{p-1} over D_p.

A cycle is thus deformed into a cycle, and a bounding cycle into a bounding cycle. It should be noted, however, that the process of deformation is not uniquely defined. Suppose that C'_p is a cycle, and that by two different deformations, over D_{p+1} and d_{p+1}, we arrive at cycles C_p and c_p of K. Then
$$D_{p+1} - d_{p+1} \to c_p - C_p$$
in some complex on K. Apply the deformation theorem to $D_{p+1} - d_{p+1}$. By (iv), we deduce that
$$C_p \sim c_p \qquad (\text{in } K).$$

† The standard proofs of this theorem only deal with the case $v = v' = 0$. The refinement of the theorem stated in (i) is easily proved. It is of importance in applications.

We can now extend the notion of homologous cycles, previously defined only for cycles of the complex K, to any pair of p-cycles lying on K. Two cycles Γ_p, Γ'_p on K are said to be homologous if $\Gamma_p - \Gamma'_p$, or a sub-division of it, bounds a chain on K. If we deform Γ_p and Γ'_p, over D_{p+1} and D'_{p+1}, into cycles Δ_p and Δ'_p of K, we have, from the deformation theorem,

$$\Delta_p \sim \Gamma_p \sim \Gamma'_p \sim \Delta'_p,$$

and hence, using the deformation theorem,

$$\Delta_p \sim \Delta'_p$$

in K. Conversely, if $\qquad \Delta_p \sim \Delta'_p \qquad (\text{in } K),$

$$\Gamma_p \sim \Delta_p \sim \Delta'_p \sim \Gamma'_p.$$

We extend, similarly, the notion of homology with division.

13·4. The deductions which can be made from the deformation theorem enable us to obtain certain invariants of the space whose points are the points of the simplexes of a complex K of class v. If two p-cycles Γ_p and Γ'_p on K, both of class $v_1 \leqslant v$, are homologous, $\Gamma_p - \Gamma'_p$ is the boundary of a chain of class v_1 on K. We denote the set of cycles of class v_1 homologous to Γ_p by $\{\Gamma_p\}_{v_1}$ or $\{\Gamma'_p\}_{v_1}$, where Γ'_p is any cycle of class v_1 homologous to Γ_p. The sets of homologous p-cycles of class v_1 form an additive group, the law of composition being given by

$$\{\Gamma_p\}_{v_1} + \{\Gamma'_p\}_{v_1} = \{\Gamma_p + \Gamma'_p\}_{v_1}.$$

The deformation theorem tells us that in any set of homologous cycles there are cycles belonging to the complex K, and that two cycles of K belong to the same set $\{\Gamma_p\}_{v_1}$ of homologous cycles if and only if they are homologous in K. It follows that the group which we have defined is isomorphic with the direct sum of the pth Betti group and the pth torsion group of K. The Betti number R_p, and the torsion coefficients $d_1, \ldots, d_{\theta_p}$, are therefore characters of the space.

The group formed by the sets $\{\Gamma_p\}_{v_1}$ of homologous cycles is, therefore, independent of v_1. But it is important, in view of the applications which we shall later make to the theory of integrals, to observe that we can confine our attention to chains and cycles of a given class v_1. We call the group formed by the sets $\{\Gamma_p\}$ the pth *homology group* of the space whose points are the points of the simplexes of K.

Certain immediate corollaries may be mentioned. If K_1 is any non-singular complex of class v_1 lying on K, such that K is a complex lying on K_1, we could determine the homology groups of our space either by K or by K_1. Consequently the Betti numbers and the torsion coefficients of K_1 are equal to those of K. Again, if we have two spaces S_1 and S_2 which can each be covered by a non-singular complex and which are in (1-1) correspondence of class $u \geqslant 0$, they have isomorphic homology groups. Two spaces in (1-1) correspondence of class v are said to be *homeomorphic* (of class v), and our conclusion is that the homology groups of homeomorphic spaces are the same. Characters of a space which are invariant under homeomorphism are called topological invariants of the space.

13·5. We must now consider some special complexes which we shall need in what follows. If H_n is any sphere of n dimensions, we can easily construct a non-singular complex K lying on H_n, with the property that every point of H_n lies on K. We find that the following properties hold:

(i) every q-simplex ($q < n$) of K lies on the boundary of an n-simplex of K;

(ii) if the n-simplexes of K are suitably oriented, their chain-sum (with coefficients $+1$) is a cycle;

(iii) the Betti numbers of K are given by

$$R_0 = R_n = 1; \; R_p = 0 \qquad (0 < p < n);$$

(iv) there are no zero divisors, of any dimension.

These properties do not characterise a sphere, but in our topological investigations we do not require any other properties of a sphere, and we may therefore use the term "sphere" to denote any complex with the properties (i), (ii), (iii) and (iv). If the complex is of class v, the sphere is said to be of class v.

Next, let us consider a complex K consisting of an n-simplex and the simplexes on its boundary. Let L denote the complex formed by the simplexes of K of dimension less than n. Now make any sub-division K' of K. The complex L is replaced by a complex L', which can be defined as follows. Let Γ'_{n-1} be the boundary of the chain-sum of the n-simplexes of K', and let Γ_{n-1} be the chain obtained from Γ'_{n-1} by reducing the coefficients modulo 2. The complex L' consists of the simplexes which have unit coefficients in Γ_{n-1}, and the simplexes on their boundaries. The following results hold:

(i) every q-simplex $(q<n)$ of K' lies on the boundary of an n-simplex;

(ii) the n-simplexes of K can be so oriented that their chain-sum is a chain Γ_n whose boundary is Γ_{n-1};

(iii) L' is a sphere H_{n-1};

(iv) K' has the Betti numbers

$$R_0 = 1; \; R_p = 0 \qquad (0<p\leqslant n);$$

(v) there are no zero divisors, of any dimensions.

If K' is any complex, and L' the sub-complex of K' defined as above, we say that $K'-L'$ is an *n-cell* if the properties (i), ..., (v) are satisfied. The chain Γ_n is then called an *oriented n-cell*. A simplex is thus a cell.

The importance of the concept of a cell lies in the following fact. Let K be any non-singular complex, and suppose the simplexes of K are grouped together in such a way that each group forms a cell, and the set of cells has the properties that (i) no two cells have a point in common; (ii) the boundary of any cell is made up of cells of the set. The set of cells then forms a complex K_1 of cells, and the analysis of § 12·2 can be applied to K_1. If we now consider any cycle Γ_p on K, we can deform it into a cycle of K, and if we then consider the part of this cycle which lies on a cell of K_1, it can be shown that we can deform

Γ_p into a cycle of K_1. Thus we can apply all the results which we have already obtained to complexes of cells. This is of importance in our investigations of the topology of manifolds.

14. **Manifolds.** The object of this summary is to find the topological characters of a Riemannian manifold which will be used in later chapters, and the purpose of §§ 12 and 13 has been to provide the means of doing this. The connection between the ideas discussed in these paragraphs and a Riemannian manifold is provided by the theorem usually called the *Covering Theorem*. This theorem states that, given any manifold of class u, there exists a complex K of class v, for any given v $(0 \leqslant v \leqslant u)$, with the property that every point of M lies on one and only one simplex of K, and every simplex of K lies on M. This important theorem was first proved, for algebraic manifolds only, by van der Waerden [11]. Later, his proof was extended to cover the case of analytic varieties by Lefschetz and Whitehead [6], and an alternative proof, valid for analytic varieties, was given by Brown and Koopman [2]. The proof of the theorem for manifolds of class u is given by Cairns [3]. The reader is also referred to a recent paper by Whitehead on the covering theorem [12]. The complex K is not, of course, unique, but the results stated in § 13·4 show that any covering complex K will serve to determine the homology groups of M.

It should be pointed out that the topological properties of a manifold are introduced on account of their importance in the theory of multiple integrals on the manifold. The field over which a p-fold integral is to be evaluated is, indeed, a p-chain. The definition which we shall give of the value of an integral over a chain will be valid if the chain is of class *one*, but in order to perform certain operations it is convenient to confine our attention to chains of class *two* at least. Of special importance is the problem of finding the relation between the values of an integral over two homologous cycles, and in this connection we make use of the result, pointed out above, that if Γ_p and Γ'_p are two cycles of class v which are homologous, $\Gamma_p - \Gamma'_p$ is the boundary of a chain of class v.

The situation is as follows. On a manifold of class u, the operations which we have to perform on chains of class v ($v \leqslant u$) can all be described in terms of chains of class v, and, if $v \geqslant 2$, they are therefore permissible in our investigations on integrals. The topological characters of the manifold do not, however, depend on the class number v, and consequently it will not be necessary, in the remainder of this summary, to specify the classes of the simplexes with which we deal.

A complex K which covers a Riemannian manifold M has certain properties not possessed by general complexes, and we now go on to describe those special properties which will be used in later chapters. We shall first see that the n-simplexes can be so oriented that their chain sum is a cycle Γ_n, and that any n-cycle on K is equal to $\lambda\Gamma_n$, where λ is an integer. We shall then introduce a dual complex, and show that an important duality relation holds for K. From this we shall go on to define the intersection of two cycles on a manifold, and obtain some new topological invariants of the manifold. Finally, we introduce the notion of the product of two manifolds, and state, without proof, certain formulae which will be of frequent use in the theory of integrals.

15. **Orientation.** We observe that, if K is any complex covering a Riemannian manifold M, every p-simplex ($p < n$) of K lies on the boundary of an n-simplex, and that every $(n-1)$-simplex of K lies on the boundary of two n-simplexes. By replacing K by a complex obtained by repeated regular sub-division, if necessary, we may suppose that each n-simplex of K lies in the domain of at least one set of local coordinates on M, that is, in a neighbourhood of M. Now M is orientable, and we can therefore find coordinate systems

$$(x^i_1, \ldots, x^i_n) \qquad (i = 1, \ldots, \alpha_n),$$

each belonging to the class of coordinate systems which orient M, such that E^i_n lies in the neighbourhood in which

$(x_1^i, \ldots, x_n^i)$ is a local coordinate system. We saw in § 13·2 how we could orient a simplex by means of a set of parameters; we therefore orient E_n^i by the parametric system $(x_1^i, \ldots, x_n^i)$. We shall always assume that the n-simplexes are oriented according to this rule.

Suppose that E_{n-1}^i lies on the boundary of E_n^j and E_n^k. Then

$$_{(n)}\eta_i^r = 0 \qquad (r \neq j, k),$$

and a simple theorem in calculus shows that

$$_{(n)}\eta_i^j + {}_{(n)}\eta_i^k = 0.$$

Hence
$$\Gamma_n = \sum_{i=1}^{\alpha_n} E_n^i \to 0.$$

Now suppose that
$$\Gamma_n' = a_i E_n^i$$

is any other n-cycle of K. Then $\Gamma_n' - a_1 \Gamma_n$ is a cycle in which at least one of the n-simplexes of K does not appear. It can be shown that if this cycle is not zero the manifold M is not connected (§ 2·1). But a Riemannian manifold is connected. Hence

$$\Gamma_n' = a_1 \Gamma_n,$$

and the nth Betti number of K is equal to *one*.

16·1. **Duality.** Let K be any complex covering a Riemannian manifold M. We can construct a complex K' by regular sub-division of K. To do this, we have to introduce a new vertex (0-simplex) into each simplex of K. Let P_p^i denote the vertex introduced in the simplex E_p^i. By the sub-division, any simplex E_p^i is replaced by a chain of K', formed from simplexes of K' which lie on E_p^i. Now it can easily be seen that any p-simplex of K' on E_p is of the form

$$\eta P_0^{i_0} \ldots P_{p-1}^{i_{p-1}} P_p^i \qquad (\eta = \pm 1),$$

where $E_{p-1}^{i_{p-1}}$ is on the boundary of $E_p^i, \ldots, E_{k-1}^{i_{k-1}}$ is on the boundary of $E_k^{i_k}$. E_p^i is therefore replaced by a chain

$$\Sigma \eta P_0^{i_0} \ldots P_{p-1}^{i_{p-1}} P_p^i,$$

where the summation is over all the p-simplexes of K' which lie on E_p^i, each having coefficient $+1$. By convention, the orientation of the simplexes involved is fixed by the orientation of E_p^i.

We now consider any $(n-p)$-simplex of K' which can be written in the form

$$\xi P_p^i P_{p+1}^{i_{p+1}} \ldots P_n^{i_n} \qquad (\xi = \pm 1),$$

where E_p^i is on the boundary of $E_{p+1}^{i_{p+1}}, \ldots, E_k^{i_k}$ is on the boundary of $E_{k+1}^{i_{k+1}}$. The orientation of this simplex is not fixed by our convention, except when $p = n$. We now fix the orientation as follows.

$$\zeta P_0^{i_0} \ldots P_{p-1}^{i_{p-1}} P_p^i P_{p+1}^{i_{p+1}} \ldots P_n^{i_n}$$

is an n-simplex of K' whose orientation has been fixed. We orient the $(n-p)$-simplex so that

$$\xi\eta\zeta = 1.$$

It can be shown that the orientation of this $(n-p)$-simplex is determined by the orientation of E_p^i, and does not depend on the particular p-simplex of K' chosen on E_p^i.

We now consider the $(n-p)$-chain of K':

$$E_{n-p}^{*i} = \Sigma\, \xi P_p^i P_{p+1}^{i_{p+1}} \ldots P_n^{i_n},$$

summed over all such $(n-p)$-simplexes of K' whose first vertex is P_p^i. The manifold M has the property that, whatever covering complex K we choose, E_{n-p}^{*i}† is an $(n-p)$-cell. The unoriented cell consists of simplexes of K' which all have P_p^i as a vertex. E_{n-p}^{*i} is the *dual* of the simplex E_p^i.

16·2. It is not difficult to show that the set of dual cells E_{n-p}^{*i} $(i = 1, \ldots, \alpha_p;\ p = 0, \ldots, n)$ form a complex K^* which

† It would be more correct to denote the dual cell by $E_{n-p,i}^{*}$, but for typographical reasons we prefer E_{n-p}^{*i}. The reader will later notice other slight departures from the strictly logical notation made for similar reasons. As a consequence, we shall occasionally have to introduce the summation sign Σ to sum over equal indices placed in the same position.

covers M. We call it the *dual complex* of K. The incidence relations of K^* are

$$E^{*i}_{n-p} \to {}_{(n-p)}\eta^{*i}_j \, E^j_{n-p-1},$$

where
$$_{(n-p)}\eta^* = (-1)^{p+1} {}_{(p+1)}\eta',$$

η' denoting the transpose of η.

The pth Betti number of K^* is

$$\begin{aligned} R^*_p &= \alpha^*_p - \rho^*_p - \rho^*_{p+1} \\ &= \alpha_{n-p} - \rho_{n-p+1} - \rho_{n-p} \\ &= R_{n-p}, \end{aligned}$$

and the pth torsion coefficients of K^* are given by the invariant factors of ${}_{(p+1)}\eta^*$, i.e. of ${}_{(n-p)}\eta$. Hence the pth torsion coefficients of K^* are equal to the $(n-p-1)$th torsion coefficients of K.

We have already seen that a complex K^* of cells which covers M will serve to determine the homology groups of M as well as a complex K of simplexes. Hence the pth Betti number of M is equal to R_p, and to $R^*_p = R_{n-p}$, and the pth torsion coefficients of M are equal to the invariant factors of ${}_{(p+1)}\eta$, and to those of ${}_{(p+1)}\eta^* = (-1)^{n-p} {}_{(n-p)}\eta'$. Hence,

(i) the pth and $(n-p)$th Betti numbers of M are equal;

(ii) the pth and $(n-p-1)$th torsion coefficients of M are the same.

These two results constitute the *Duality Theorem* of Poincaré. We can further prove that the torsion coefficients of dimension 0, $n-1$, n are all unity.

17·1. **Intersections.** The introduction of the dual complex K^* on a manifold M enables us to develop the idea of the intersection of two cycles. We first consider the intersection of a chain or cycle of K with a chain or cycle of K^*, and then pass on to use the results so obtained to define the intersection of cycles on M.

Let E_p^i be any p-simplex of the complex K, and E_q^{*j} a q-cell of the dual complex K^*. It can be seen that if $p+q<n$, E_p^i and E_q^{*j} have no point in common. We then write

$$E_p^i \,.\, E_q^{*j} = 0.$$

If $p+q = n$, the cells have no point in common unless $i = j$, in which case they meet in P_p^i. We write

$$E_p^i \,.\, E_q^{*j} = 0 \qquad (i \neq j),$$

and

$$E_p^i \,.\, E_q^{*i} = P_p^i.$$

If $p+q>n$, E_p^i and E_q^{*j} either have no point in common or meet in a number of simplexes of the form

$$\pm P_{n-q}^j P_{n-q+1}^{k_{n-q+1}} \dots P_{p-1}^{k_{p-1}} P_p^i$$

of K'. The orientation of such a $(p+q-n)$-simplex has not been fixed as yet by our conventions. We now fix it as follows. There exist simplexes

$$\eta P_0^{k_0} \dots P_{n-q-1}^{k_{n-q-1}} P_{n-q}^j P_{n-q+1}^{k_{n-q+1}} \dots P_p^i$$

and

$$\zeta P_{n-q}^j P_{n-q+1}^{k_{n-q+1}} \dots P_p^i P_{p+1}^{k_{p+1}} \dots P_n^{k_n}$$

of K' lying on E_p^i and E_{n-q}^{*j} respectively, and an n-simplex

$$\zeta P_0^{k_0} \dots P_{n-q-1}^{k_{n-q-1}} P_{n-q}^j P_{n-q+1}^{k_{n-q+1}} \dots P_{p-1}^{k_{p-1}} P_p^i P_{p+1}^{k_{p+1}} \dots P_n^{k_n}$$

of K', whose orientations have all been fixed by our conventions. If we orient the $(p+q-n)$-simplex by writing it as

$$\xi\eta\zeta P_{n-q}^j P_{n-q+1}^{k_{n-q+1}} \dots P_{p-1}^{k_{p-1}} P_p^i,$$

it can be shown that this orientation depends only on the orientation of E_p^i, E_q^{*j}. We now write

$$E_p^i \,.\, E_q^{*j} = \Sigma \xi\eta\zeta\, P_{n-q}^j P_{n-q+1}^{k_{n-q+1}} \dots P_{p-1}^{k_{p-1}} P_p^i,$$

summed over the simplexes common to E_p^i, E_q^{*j}. It can be shown that $E_p^i \,.\, E_q^{*j}$ is a $(p+q-n)$-cell, or else zero.

The oriented cell $E_p^i \,.\, E_q^{*j}$ is defined as the *intersection* of E_p^i and E_q^{*j}. From the definition, it follows that

$$(-E_p^i) \,.\, E_q^{*j} = -E_p^i \,.\, E_q^{*j} = E_p^i \,.\, (-E_q^{*j}).$$

Let
$$C_p = a_i E_p^i$$
and
$$C_q^* = b_j E_q^{*j}$$
be chains of K and K^*, respectively. We define the intersection $C_p \,.\, C_q^*$ by the formula

$$C_p \,.\, C_q^* = a_i b_j E_p^i \,.\, E_q^{*j}.$$

This is a $(p+q-n)$-chain of K'. By simple calculations we can show that if

$$C_p \to C_{p-1}, \quad C_q^* \to C_{q-1}^*,$$

then
$$C_p \,.\, C_q^* \to C_p \,.\, C_{q-1}^* + (-1)^{n-q} C_{p-1} \,.\, C_q^*.$$

In particular, if C_p and C_q^* are cycles, $C_p \,.\, C_q^*$ is a cycle. Moreover, if

$$C_p \approx 0,$$

say
$$C_{p+1} \to \lambda C_p,$$
and
$$C_q^* \to 0,$$
then
$$C_{p+1} \,.\, C_q^* \to (-1)^{n-q} \lambda C_p \,.\, C_q^*,$$
that is
$$C_p \,.\, C_q^* \approx 0.$$

A similar result holds when

$$C_p \to 0, \quad C_q^* \approx 0.$$

It follows that if
$$C_p \approx C_p' \quad \text{(in } K\text{)},$$
and
$$C_q^* \approx C_q^{*\prime} \quad \text{(in } K^*\text{)},$$
then
$$C_p \,.\, C_q^* \approx C_p' \,.\, C_q^{*\prime} \quad \text{(in } K'\text{)}.$$

17·2. We are now in a position to define the intersection of any two cycles Γ_p and Γ_q on the manifold M. Practically nothing can be said about the set of points common to two cycles Γ_p and Γ_q, but in any case we are not concerned with this set of points. We can, however, define an invariant of the sets $\{\Gamma_p\}$ and $\{\Gamma_q\}$ of cycles homologous to Γ_p and Γ_q, having most of the properties which we commonly associate with intersections.

In $\{\Gamma_p\}$ there exists a cycle C_p of K, and in $\{\Gamma_q\}$ there exists a cycle C_q^* of K^*. The cycle C_p is determined to within a bounding cycle of K, and C_q^* is determined to within a bounding cycle of K^*. The cycle $C_p \,.\, C_q^*$ of K' is therefore determined by $\{\Gamma_p\}$ and $\{\Gamma_q\}$ to within a bounding cycle of K'. Thus $\{\Gamma_p\}$ and $\{\Gamma_q\}$ determine, by means of the complexes K and K^*, a set of homologous cycles $\{C_p \,.\, C_q^*\}$, of dimension $p+q-n$. A theorem analogous to the deformation theorem can then be proved to show that the set of cycles $\{C_p \,.\, C_q^*\}$ is independent of K and K^*. Any cycle of this set is then defined to be the intersection of Γ_p and Γ_q, and is denoted by $\Gamma_p \,.\, \Gamma_q$.

It should be observed that this definition of intersection coincides with our intuitive notion of an intersection in several important cases. If Γ_p and Γ_q are written as chain-sums of simplexes

$$\Gamma_p = a_i e_p^i$$

and

$$\Gamma_q = b_i f_q^i,$$

and if the simplexes e_p^i, f_q^j either have no points in common, or meet in an $(p+q-n)$-cell, which we denote by $e_p^i \,.\, f_q^j$ (with a suitable orientation), then the chain

$$a_i b_j \, e_p^i \,.\, f_q^j$$

is a cycle of the set $\{\Gamma_p \,.\, \Gamma_q\}$. Of particular importance in later chapters is the case in which M is the Riemannian of an algebraic variety V_m of m complex dimensions. The dimension of M is $2m$. An algebraic variety of dimension r on V_m defines a $2r$-cycle Γ_{2r} on M. Let A and B be two algebraic varieties on V_m of dimensions a and b, determining cycles Γ_{2a}, Γ_{2b} on M, and suppose that the geometrical intersection of A and B is a variety of dimension $a+b-m$ (counted with proper multiplicity). This intersection defines a cycle $\Gamma_{2(a+b-m)}$ on M. We can show that

$$\Gamma_{2a} \,.\, \Gamma_{2b} \sim \pm \Gamma_{2(a+b-m)}.$$

17·3. The intersections of cycles on M have the following properties:

(i) $$(a\Gamma_p+b\Gamma'_p)\,.\,\Gamma_q \sim a\Gamma_p\,.\,\Gamma_q+b\Gamma'_p\,.\,\Gamma_q;$$

(ii) $$\Gamma_p\,.\,(a\Gamma_q+b\Gamma'_q) \sim a\Gamma_p\,.\,\Gamma_q+b\Gamma_p\,.\,\Gamma'_q;$$

(iii) if $$\Gamma_p \sim 0, \quad \text{or} \quad \Gamma_q \sim 0,$$

then $$\Gamma_p\,.\,\Gamma_q \sim 0;$$

(iv) $$\Gamma_p\,.\,\Gamma_q \sim (-1)^{(n-p)(n-q)}\,\Gamma_q\,.\,\Gamma_p.$$

We can extend the notion of the intersection of two cycles, and define the intersection of three or more cycles. Let Γ_p, Γ_q, Γ_r be three cycles on M. We define the triple intersection to be

$$\Gamma_p\,.\,\Gamma_q\,.\,\Gamma_r \sim (\Gamma_p\,.\,\Gamma_q)\,.\,\Gamma_r.$$

It can then be shown that

$$\Gamma_p\,.\,\Gamma_q\,.\,\Gamma_r \sim \Gamma_p\,.\,(\Gamma_q\,.\,\Gamma_r).$$

17·4. When we consider the intersection of two cycles Γ_p and Γ_q, where $p+q=n$, we obtain certain numerical invariants of the cycles. We first return to the considerations of § 17·1, and consider chains of K of dimension p and chains of K^* of dimension $n-p$.

If $$C_p = a_i E^i_p$$

is a p-chain of K, and

$$C^*_{n-p} = b_i E^{*i}_{n-p}$$

is an $(n-p)$-chain of K^*, then

$$C_p\,.\,C^*_{n-p} = \sum_{i=1}^{\alpha_p} a_i b_i P^i_p.$$

We now define the *intersection number* (or *Kronecker index*) of C_p and C^*_{n-p} to be

$$(C_p\,.\,C^*_{n-p}) = \sum_{i=1}^{\alpha_p} a_i b_i.$$

We observe that $$(E^i_p\,.\,E^{*j}_{n-p}) = \delta^i_j,$$

that is, in matrix notation,

$$\|(E_p^i \,.\, E_{n-p}^{*j})\| = \mathbf{I}_{\alpha_p}.$$

We consider the change of base for the chains of K of dimension $(p+1)$, p, $(p-1)$,

$$\begin{aligned} C_{p+1}^i &= \alpha_j^i E_{p+1}^j, \\ C_p^i &= \beta_j^i E_p^j, \\ C_{p-1}^i &= \gamma_j^i E_{p-1}^j, \end{aligned}$$

used to obtain the canonical forms (8) for the incidence relations for the chains of K. We now make the changes of base,

$$\begin{aligned} C_{n-p+1}^{*i} &= \gamma_j^{*i} E_{n-p+1}^{*j}, \\ C_{n-p}^{*i} &= \beta_j^{*i} E_{n-p}^{*j}, \\ C_{n-p-1}^{*i} &= \alpha_j^{*i} E_{n-p-1}^{*j}, \end{aligned}$$

for the chains of K^*, where

$$\boldsymbol{\alpha}^*\boldsymbol{\alpha}' = \mathbf{I}_{\alpha_{p+1}}, \quad \boldsymbol{\beta}^*\boldsymbol{\beta}' = \mathbf{I}_{\alpha_p}, \quad \boldsymbol{\gamma}^*\boldsymbol{\gamma}' = \mathbf{I}_{\alpha_{p-1}}.$$

We obtain the incidence relations (in which we do *not* use the summation convention),

$$\begin{aligned} & C_{n-p+1}^{*i} \to (-1)^p e_i\, C_{n-p}^{*\rho_{p+1}+i} && (i=1, \ldots, \rho_p); \\ & C_{n-p+1}^{*j} \to 0 && (j=\rho_p+1, \ldots, \alpha_{p-1}); \\ & C_{n-p}^{*i} \to (-1)^{p-1} d_i\, C_{n-p-1}^{*i} && (i=1, \ldots, \rho_{p+1}), \\ & C_{n-p}^{*i} \to 0 && (i=\rho_{p+1}+1, \ldots, \rho_{p+1}+\rho_p), \\ & C_{n-p}^{*i} \to 0 && (i=\rho_{p+1}+\rho_p+1, \ldots, \alpha_p). \end{aligned}$$

Also,

$$\begin{aligned} \|(C_p^i \,.\, C_{n-p}^{*j})\| &= \boldsymbol{\beta}\, \|(E_p^i \,.\, E_{n-p}^{*j})\|\, \boldsymbol{\beta}^{*\prime} \\ &= \boldsymbol{\beta} \mathbf{I}_{\alpha_p} \boldsymbol{\beta}^{-1} \\ &= \mathbf{I}_{\alpha_p}. \end{aligned}$$

Now

$$\Gamma_p^i = C_p^{\rho_{p+1}+\rho_p+i} \qquad (i=1, \ldots, R_p)$$

form a set of R_p p-cycles of K, no linear combination of which is homologous to zero, and

$$\Gamma_{n-p}^{*i} = C_{n-p}^{*\rho_{p+1}+\rho_p+i} \qquad (i=1, \ldots, R_{n-p})$$

form a set of R_{n-p} $(n-p)$-cycles of K^*, no linear combination of which is homologous to zero. The results just found show that

(i) $$\|(\Gamma_p^i \,.\, \Gamma_{n-p}^{*j})\| = \mathbf{I}_{R_p};$$

(ii) if $\Gamma_p \approx 0$, $(\Gamma_p \,.\, \Gamma_{n-p}^{*i}) = 0 \quad (i=1, \ldots, R_{n-p})$;

(iii) if $\Gamma_{n-p}^* \approx 0$, $(\Gamma_p^i \,.\, \Gamma_{n-p}^*) = 0 \quad (i=1, \ldots, R_p)$;

(iv) if $\Gamma_p \approx 0$, $\Gamma_{n-p}^* \approx 0$, then $(\Gamma_p \,.\, \Gamma_{n-p}^*) = 0$.

17·5. We can now define the intersection number $(\Gamma_p \,.\, \Gamma_{n-p})$ of two cycles Γ_p and Γ_{n-p} on M. If we deform Γ_p into a cycle C_p of K, we have

$$C_p = \sum_{i=1}^{\rho_{p+1}} a_i C_p^i + \sum_{i=1}^{R_p} b_i \Gamma_p^i.$$

The method of obtaining C_p from Γ_p is not unique, but all cycles of K obtained by deformation from Γ_p are homologous in K. Hence the numbers $b_1, \ldots, b_{R_p}$ are uniquely determined, and we have

$$\Gamma_p \approx \sum_{i=1}^{R_p} b_i \Gamma_p^i.$$

In the same way, if the cycle Γ_{n-p} is deformed into the cycle C_{n-p}^* of K^*,

$$C_{n-p}^* = \sum_{i=\rho_{p+1}+1}^{\rho_{p+1}+\rho_p} a_i' C_{n-p}^{*i} + \sum_{i=1}^{R_{n-p}} b_i' \Gamma_{n-p}^{*i},$$

where $b_1', \ldots, b_{R_{n-p}}'$ are uniquely determined, and

$$\Gamma_{n-p} \approx \sum_{i=1}^{R_{n-p}} b_i' \Gamma_{n-p}^{*i}.$$

We define the intersection number $(\Gamma_p \,.\, \Gamma_{n-p})$ by the equation

$$(\Gamma_p \,.\, \Gamma_{n-p}) = (C_p \,.\, C_{n-p}^*) = \sum_{i=1}^{R_p} b_i b_i'.$$

As it is proved that the definition of the intersection of two cycles is independent of the covering complexes K and K^*, so it can also be proved that the intersection number does not depend on K or K^*. Moreover,

$$(\Gamma_p \cdot \Gamma_{n-p}) = (-1)^{p(n-p)} (\Gamma_{n-p} \cdot \Gamma_p).$$

17·6. The cycles Γ_p^i $(i = 1, \ldots, R_p)$ have the property that any p-cycle on M is related to them by a homology with division, of the form

$$\Gamma_p \approx a_i \Gamma_p^i,$$

where the coefficients a_i are integers. Any set of R_p p-cycles of M with this property form a *fundamental base* for the p-cycles of M. It is easily verified that the cycles Δ_p^i $(i = 1, \ldots, R_p)$ form a fundamental base if and only if

$$\Delta_p^i \approx a_j^i \Gamma_p^j,$$

where (a_j^i) is a unimodular matrix of integers.

A set of cycles $\Gamma_p'^i$ $(i = 1, \ldots, R_p)$ is said to form a *base* for the p-cycles of M if

$$\Gamma_p'^i \approx \alpha_j^i \Gamma_p^j,$$

where (α_j^i) is a non-singular matrix of integers (not necessarily unimodular). If Γ_p is any p-cycle of M,

$$\Gamma_p \approx a_i \Gamma_p^i.$$

Since $$|\alpha_j^i| \, \Gamma_p^i \approx \beta_j^i \Gamma_p'^j,$$

where $$\beta_j^i \alpha_k^j = |\alpha_j^i| \, \delta_k^i,$$

we have $$|\alpha_j^i| \, \Gamma_p \approx a_i \beta_j^i \Gamma_p'^j,$$

a relation which we can write in the form

$$\Gamma_p \approx b_i \Gamma_p'^i,$$

where the coefficients b_i are rational numbers. When we are concerned only with homology with division, we may introduce the idea of rational multiples of a cycle, and if Δ_p is any cycle in this generalised sense, there exists an integer λ such that $\lambda \Delta_p$ is a cycle according to our original definition.

The cycles Γ^{*i}_{n-p} form a fundamental base for the $(n-p)$-cycles of M, and a set of cycles Δ^i_{n-p} $(i=1, \ldots, R_{n-p})$ forms a base for the $(n-p)$-cycles if and only if

$$\Delta^i_p \approx \gamma^i_j \Gamma^{*j}_{n-p},$$

where (γ^i_j) is a non-singular matrix of integers. The base is fundamental when (γ^i_j) is unimodular.

Now $$(\Delta^i_p \,.\, \Delta^j_{n-p}) = \alpha^i_h(\Gamma^h_p \,.\, \Gamma^{*k}_{n-p})\,\gamma^j_k,$$

that is $$\|(\Delta^i_p \,.\, \Delta^j_{n-p})\| = \boldsymbol{\alpha}\mathbf{I}_{R_p}\boldsymbol{\gamma}' = \boldsymbol{\alpha}\boldsymbol{\gamma}'.$$

This is a non-singular matrix. It follows easily that if Δ^i_p $(i=1, \ldots, R_p)$ is any base for the p-cycles of M, the $(n-p)$-cycles Δ^i_{n-p} $(i=1, \ldots, R_{n-p})$ form a base for the $(n-p)$-cycles of M if and only if

$$\|(\Delta^i_p \,.\, \Delta^j_{n-p})\|$$

is a non-singular matrix.

If Δ^i_p $(i=1, \ldots, R_p)$ is a base for the p-cycles on M, and Γ_p is any p-cycle of M,

$$\lambda\Gamma_p \approx a_i \Delta^i_p,$$

where $\lambda, a_1, \ldots, a_{R_p}$ are integers. Let Δ^i_{n-p} $(i=1, \ldots, R_{n-p})$ be a base for the $(n-p)$-cycles of M. Then

$$\lambda(\Gamma_p \,.\, \Delta^i_{n-p}) = a_j(\Delta^j_p \,.\, \Delta^i_{n-p}) \qquad (i=1, \ldots, R_{n-p}).$$

Since the matrix $\|(\Delta^i_p \,.\, \Delta^j_{n-p})\|$ is non-singular, these equations are sufficient to determine the ratios of $\lambda : a_1 : \ldots : a_{R_p}$ when the intersection numbers $(\Gamma_p \,.\, \Delta^i_{n-p})$ are given, and it follows that the necessary and sufficient condition that $\Gamma_p \approx 0$ is that

$$(\Gamma_p \,.\, \Delta^i_{n-p}) = 0 \qquad (i=1, \ldots, R_{n-p}),$$

where Δ^i_{n-p} $(i=1, \ldots, R_{n-p})$ is a base for the $(n-p)$-cycles.

We shall often find it convenient to choose bases Δ^i_p

$(i=1, \ldots, R_p;\ p=0, \ldots, n)$ once and for all for the cycles of M. We then get a set of intersection matrices

$$\mathbf{a}_p = \| (\Delta_p^i \cdot \Delta_{n-p}^j) \| \qquad (p=0, \ldots, n),$$

where

$$\mathbf{a}_p = (-1)^{p(n-p)} \mathbf{a}'_{n-p}.$$

17·7. In the investigations on the properties of algebraic varieties which we shall make in Chapter IV, we shall find the case in which the dimension n of the manifold M is even, say $n = 2m$, of particular importance. In this case, when we consider a base Δ_m^i $(i=1, \ldots, R_m)$ for the m-cycles, the intersection matrix

$$\mathbf{a}_m = \| (\Delta_m^i \cdot \Delta_m^j) \|$$

satisfies the equation

$$\mathbf{a}_m = (-1)^m \mathbf{a}'_m.$$

If we make a change of base, given by

$$\bar{\Delta}_m^i \approx p_j^i \Delta_m^j,$$

the intersection matrix is transformed into

$$\bar{\mathbf{a}}_m = \mathbf{p}\mathbf{a}_m\mathbf{p}'.$$

There are two cases to consider. Case (a): m odd. Then $\mathbf{a}_m$ is skew. But it is non-singular. Therefore R_m must be even. We can easily show that a fundamental base can be chosen so that

$$\bar{\mathbf{a}}_m = \begin{pmatrix} \mathbf{0} & \mathbf{I}_{\frac{1}{2}R_m} \\ -\mathbf{I}_{\frac{1}{2}R_m} & \mathbf{0} \end{pmatrix}.$$

Case (b): m even. Then $\mathbf{a}_m$ is symmetrical. In this case we can choose $\mathbf{p}$ so that $\bar{\mathbf{a}}_m$ is a diagonal matrix. The matrix $\mathbf{p}$ is not uniquely defined by this property. But it is a well-known theorem in algebra that, whatever $\mathbf{p}$ is chosen to reduce $\mathbf{a}_m$ to diagonal form, the number of elements on the diagonal of the reduced forms which are positive is always the same. This number is called the *signature* of $\mathbf{a}_m$. It is a topological invariant of the manifold.

18·1. **Product manifolds.** Let S_1 and S_2 be two sets of elements. From these we can define a new set, whose elements are obtained by associating any element of S_1 with any element of S_2. We denote this new set by $S_1 \times S_2$ and call it the product of S_1 by S_2.

Let
$$x_i = f_i(u_1, \ldots, u_p) \qquad (i = 1, \ldots, N)$$
be any p-simplex of class v_1 in the space $(x_1, \ldots, x_N)$, and let
$$x'_i = g_i(v_1, \ldots, v_q) \qquad (i = 1, \ldots, N')$$
be a q-simplex in the space $(x'_1, \ldots, x'_{N'})$, of class v_2. The product of the two sets of points can be represented in the space $(x_1, \ldots, x_N;\, x'_1, \ldots, x'_{N'})$ by means of the equations
$$x_i = f_i(u_1, \ldots, u_p) \qquad (i = 1, \ldots, N),$$
$$x'_i = g_i(v_1, \ldots, v_q) \qquad (i = 1, \ldots, N').$$
It can be shown that this product is a $(p+q)$-cell, whose class is the smaller of v_1 and v_2. The cells on the boundary of this $(p+q)$-cell are either the products of the p-simplex by the simplexes on the boundary of the q-simplex, or the products of the simplexes on the boundary of the p-simplex by the q-simplex, or else the products of simplexes on the boundary of each.

If we orient the p-simplex and the q-simplex, and denote the oriented simplexes by E_p, E'_q, we have incidence relations
$$E_p \to \eta_i E^i_{p-1},$$
$$E'_q \to \zeta_i E'^i_{q-1}.$$
Using a suitable convention to orient the product of the simplexes, it can be shown that
$$E_p \times E'_q \to \eta_i E^i_{p-1} \times E'_q + (-1)^p \zeta_i E_p \times E'^i_{q-1}.$$

18·2. Let M_1 be a manifold of n_1 dimensions, covered by a complex K_1, which lies in the space $(x_1, \ldots, x_{N_1})$, and let M_2 be a manifold of n_2 dimensions, covered by a complex K_2, which lies in the space $(y_1, \ldots, y_{N_2})$. The product $M_1 \times M_2$ is a manifold of (n_1+n_2) dimensions, which can be represented as a locus in $(x_1, \ldots, x_{N_1}; y_1, \ldots, y_{N_2})$. It is covered by the complex $K_1 \times K_2$, whose cells are the products of simplexes of K_1 by simplexes of K_2. The incidence relations for the complex $K_1 \times K_2$ can be determined from the incidence relations of K_1 and K_2.

If C_p is a p-chain of M_1,

$$C_p = a_i E_p^i,$$

and if D_q is a q-chain of M_2,

$$D_q = b_i F_q^i,$$

we define the product chain $C_p \times D_q$ to be

$$C_p \times D_q = a_i b_j E_p^i \times F_q^j.$$

By algebraic calculation we can show that if C_p and D_q are cycles, so is $C_p \times D_q$, and that, if Γ_p^i $(i=1, \ldots, R_p)$ is a base for the p-cycles of M_1, and Δ_q^i $(i=1, \ldots, S_q)$ is a base for the q-cycles of M_2, the cycles

$$\Gamma_p^i \times \Delta_{r-p}^j \quad (i=1, \ldots, R_p;\ j=1, \ldots, S_{r-p};\ p=0, \ldots, r)$$

form a base for the r-cycles of $M_1 \times M_2$. The rth Betti number of $M_1 \times M_2$ is therefore

$$\sum_{p=0}^{r} R_p S_{r-p}.$$

The intersection theory on $M_1 \times M_2$ follows from the formula

$$E_p^i \times F_q^j \,.\, E_r^{*k} \times F_s^{*l} = (-1)^{(n_1-p)(n_2-s)}\, E_p^i \,.\, E_r^{*k} \times F_q^j \,.\, F_s^{*l}.$$

From this we obtain the equation

$$\Gamma_p \times \Delta_q \,.\, \Gamma_r \times \Delta_s \approx (-1)^{(n_1-p)(n_2-s)}\, \Gamma_p \,.\, \Gamma_r \times \Delta_q \,.\, \Delta_s.$$

18·3. The case which will be of most frequent use in applications is that in which M_1 and M_2 are manifolds of n dimensions of class u, which are in (1-1) correspondence of class u. We then suppose that, in the correspondence, K_1 corresponds to K_2.

Without loss of generality, we may suppose that M_1 and M_2 lie in spaces $(x_1, \ldots, x_N)$ and $(y_1, \ldots, y_N)$ of N dimensions and that the correspondence is given by the equations $x_i = y_i$ $(i = 1, \ldots, N)$.

The product manifold $M_1 \times M_2$ is then represented as a locus of $2n$ dimensions in $(x_1, \ldots, x_N; y_1, \ldots, y_N)$. On it there is a set of points defined by the equations

$$x_i = y_i \qquad (i = 1, \ldots, N),$$

which is the image of class u of M_1 or M_2. It is therefore a cycle, which we call the *diagonal cycle* on $M_1 \times M_2$, and which we denote by Γ. Further, in this set of points there is a cycle γ_p which is the image of a cycle Γ_p of M_1. It is important to find homologies for Γ and γ_p on $M_1 \times M_2$, which express them in terms of a base for the cycles of $M_1 \times M_2$. The following formulae can be proved.

Let Γ_p^i $(i = 1, \ldots, R_p)$ be a base for the p-cycles of M $(p = 0, \ldots, n)$, and let Δ_p^i be the cycle of M which corresponds to Γ_p^i in the homeomorphism which connects M_1 and M_2. We denote the intersection matrix $\| (\Gamma_p^i \cdot \Gamma_{n-p}^j) \|$ by $\mathbf{a}_p$, and the matrix whose element in the ith row and jth column is $(\Gamma_p \cdot \Gamma_q^i \cdot \Gamma_{2n-p-q}^j)$ by $\mathbf{a}_q(\Gamma_p)$. Then, on $M_1 \times M_2$,

$$\Gamma \approx \sum_{p=0}^{n} \epsilon_{ij}^{p} \Gamma_p^i \times \Delta_{n-p}^j, \tag{9}$$

and
$$\gamma_p \approx \sum_q \lambda_{ij}^q(\Gamma_p)\, \Gamma_{p+q-n}^i \times \Delta_{n-q}^j, \tag{10}$$

where
$$\boldsymbol{\epsilon}_p = (\mathbf{a}_p')^{-1}, \tag{11}$$

and
$$[\boldsymbol{\lambda}^q(\Gamma_p)]' = (-1)^{(n-q)(n-p)}\, \mathbf{a}_q^{-1} \cdot \mathbf{a}_q(\Gamma_p) \cdot \mathbf{a}_{p+q-n}^{-1}. \tag{12}$$

REFERENCES

1. P. ALEXANDROFF and H. HOPF. *Topologie* (Berlin), 1935.
2. A. B. BROWN and B. O. KOOPMAN. *Transactions of the American Mathematical Society*, 34 (1932), 231.
3. S. S. CAIRNS. *Annals of Mathematics*, 35 (1934), 579.
4. L. P. EISENHART. *Riemannian Geometry* (Princeton), 1926.
5. S. LEFSCHETZ. *Topology* (New York), 1930.
6. S. LEFSCHETZ and J. H. C. WHITEHEAD. *Transactions of the American Mathematical Society*, 35 (1933), 510.
7. M. MORSE. *The Calculus of Variations in the Large* (New York), 1934.
8. H. SEIFERT and W. THRELFALL. *Lehrbuch der Topologie* (Leipzig), 1934.
9. O. VEBLEN. *Invariants of Quadratic Differential Forms* (Cambridge), 1927.
10. O. VEBLEN and J. H. C. WHITEHEAD. *Foundations of Differential Geometry* (Cambridge), 1932.
11. B. L. VAN DER WAERDEN. *Mathematische Annalen*, 102 (1931), 337.
12. J. H. C. WHITEHEAD. *Annals of Mathematics*, 41 (1940).
13. H. WHITNEY. *Annals of Mathematics*, 37 (1936), 865.

Chapter II

INTEGRALS AND THEIR PERIODS

In this chapter we study those properties of multiple integrals on a manifold which lead up to the introduction of harmonic integrals. Our starting-point is the classical definition of a p-fold integral,

$$\int_{\mathscr{D}} A\, du^1 \dots du^p,$$

of a continuous function $A = A(u_1, \dots, u_p)$ of p variables, defined over a domain $\mathscr{D}$ of the number space $(u_1, \dots, u_p)$, as given, for instance, by Goursat [1]. Goursat's definition does not take explicit account of the orientation of a domain of integration, although ideas of orientation are implicit in the development of the theory, and our first task is therefore to pass from Goursat's definition to the notion of integration over an oriented p-cell.

19·1. **Multiple integrals.** We begin with the case in which the domain $\mathscr{D}$ of Goursat's definition is a p-simplex in the space $(u_1, \dots, u_p)$. When we regard this as an unoriented simplex we denote it by $\mathscr{E}_p$, and when we have assigned an orientation we denote the oriented simplex by E_p. Goursat provides a definition of

$$\int_{\mathscr{E}_p} A\, du^1 \dots du^p.$$

The orientation of the simplex selects one of the two classes of like parametric systems which are valid in $\mathscr{E}_p$ and the definition of the integral over E_p depends on whether $(u_1, \dots, u_p)$ is a parametric system concordant with the orientation of E_p or not. If $(u_1, \dots, u_p)$ is concordant with the orientation, we make the definition

$$\int_{E_p} A\, du^1 \dots du^p = \int_{\mathscr{E}_p} A\, du^1 \dots du^p,$$

and, if it is not, our definition is

$$\int_{E_p} A\,du^1 \dots du^p = -\int_{\mathscr{E}_p} A\,du^1 \dots du^p.$$

Thus
$$\int_{-E_p} A\,du^1 \dots du^p = -\int_{E_p} A\,du^1 \dots du^p.$$

We now make a transformation of class $v \geqslant 1$ which is (1-1) without exception, transforming $\mathscr{E}_p$ into a simplex $\overline{\mathscr{E}}_p$ of the number space $(\overline{u}_1, \dots, \overline{u}_p)$, given by the equations

$$\overline{u}_i = \overline{u}_i(u_1, \dots, u_p) \qquad (i = 1, \dots, p),$$

$$u_i = u_i(\overline{u}_1, \dots, \overline{u}_p) \qquad (i = 1, \dots, p),$$

where
$$J = \left| \frac{\partial u_i}{\partial \overline{u}_j} \right|$$

is different from zero at all points of $\overline{\mathscr{E}}_p$. The oriented simplex E_p on $\mathscr{E}_p$ defines uniquely an oriented simplex $\overline{E}_p$ on $\overline{\mathscr{E}}_p$. We consider the case in which $(u_1, \dots, u_p)$ is concordant with the orientation of E_p. By the theorem proved by Goursat,

$$\int_{\mathscr{E}_p} A\,du^1 \dots du^p = \int_{\overline{\mathscr{E}}_p} \overline{A}\,|\,J\,|\,d\overline{u}^1 \dots d\overline{u}^p,$$

where $|\,J\,|$ denotes the absolute value of J, and

$$\overline{A}(\overline{u}_1, \dots, \overline{u}_p) = A(u_1, \dots, u_p).$$

Hence
$$\int_{E_p} A\,du^1 \dots du^p = \eta \int_{\overline{E}_p} \overline{A}\,|\,J\,|\,d\overline{u}^1 \dots d\overline{u}^p,$$

where $\eta = +1$, or -1, according as $(\overline{u}_1, \dots, \overline{u}_p)$ is concordant with the orientation of $\overline{E}_p$, or not. But $(\overline{u}_1, \dots, \overline{u}_p)$ is concordant with the orientation of $\overline{E}_p$, or not, according as J is positive or negative, and hence we must have

$$\int_{E_p} A\,du^1 \dots du^p = \int_{\overline{E}_p} \overline{A} J\,d\overline{u}^1 \dots d\overline{u}^p$$

in both cases. Similar reasoning holds good when $(u_1, \dots, u_p)$

is not concordant with the orientation of E_p, and we have, in all cases,

$$\int_{E_p} A\,du^1 \ldots du^p = \int_{\bar{E}_p} \bar{A} J\,d\bar{u}^1 \ldots d\bar{u}^p.$$

19·2. We now define a p-fold integral in a simplicial region D of an n-dimensional space, or of an n-dimensional manifold. Let $(x_1, \ldots, x_n)$ be a coordinate system valid in D, and let $A_{i_1 \ldots i_p}$ be a continuous function of $(x_1, \ldots, x_n)$ in D, for values of the indices $i_1, \ldots, i_p$ equal to $1, \ldots, n$. We say that a p-fold integral in D is given by the expression

$$\int A_{i_1 \ldots i_p}\,dx^{i_1} \ldots dx^{i_p}$$

(where the summations are from 1 to n) in the coordinate system $(x_1, \ldots, x_n)$. Let E_p be an oriented p-simplex, of class *one* at least, in D, given by the differentiable equations

$$x_i = x_i(u_1, \ldots, u_p) \qquad (i = 1, \ldots, n),$$

where $(u_1, \ldots, u_p)$ are parameters of E_p concordant with its orientation, giving E_p as the image of an oriented p-simplex E'_p in the p-space $(u_1, \ldots, u_p)$. The simplex E_p need not be non-singular. The value of the integral over E_p is denoted by

$$\int_{E_p} A_{i_1 \ldots i_p} dx^{i_1} \ldots dx^{i_p}$$

and is defined by the equation

$$\int_{E_p} A_{i_1 \ldots i_p}\,dx^{i_1} \ldots dx^{i_p} = \int_{E'_p} A'_{i_1 \ldots i_p} \frac{\partial(x_{i_1}, \ldots, x_{i_p})}{\partial(u_1, \ldots, u_p)}\,du^1 \ldots du^p,$$

where $$A'_{i_1 \ldots i_p} = A_{i_1 \ldots i_p}\{x_1(u), \ldots, x_n(u)\}.$$

The value of the integral over E_p is independent of the parametric system $(u_1, \ldots, u_p)$ chosen on E_p. If $(\bar{u}_1, \ldots, \bar{u}_p)$ is any other allowable parametric system in E_p, we have, by § 19·1,

$$\int_{E'_p} A'_{i_1\dots i_p}\frac{\partial(x_{i_1},\dots,x_{i_p})}{\partial(u_1,\dots,u_p)}du^1\dots du^p$$

$$=\int_{\bar{E}'_p} \bar{A}'_{i_1\dots i_p}\frac{\partial(x_{i_1},\dots,x_{i_p})}{\partial(u_1,\dots,u_p)}\frac{\partial(u_1,\dots,u_p)}{\partial(\bar{u}_1,\dots,\bar{u}_p)}d\bar{u}^1\dots d\bar{u}^p$$

$$=\int_{\bar{E}'_p} \bar{A}'_{i_1\dots i_p}\frac{\partial(x_{i_1},\dots,x_{i_p})}{\partial(\bar{u}_1,\dots,\bar{u}_p)}d\bar{u}^1\dots d\bar{u}^p,$$

where $$\bar{A}'_{i_1\dots i_p}=A_{i_1\dots i_p}\{x_1(\bar{u}),\dots,x_n(\bar{u})\},$$

and $\bar{E}'_p$ is the p-simplex in $(\bar{u}_1,\dots,\bar{u}_p)$ corresponding to E_p.

The value of the integral over any p-simplex depends on the coefficients $A_{i_1\dots i_p}$, and on the Jacobians

$$\frac{\partial(x_{i_1},\dots,x_{i_p})}{\partial(u_1,\dots,u_p)}.$$

It will be observed, however, that the value of the integral does not depend on the individual coefficients, but depends only on the combinations $\Sigma \pm A_{i_1\dots i_p}$ of the coefficients having the same suffixes written in different orders, the positive sign being attached to the terms whose suffixes are even derangements of a certain order, and the negative sign to the remaining terms. There is therefore no loss of generality if we impose the condition that the coefficients are skew-symmetric in the suffixes, that is,

$$A_{i_1\dots i_{a-1}i_a i_{a+1}\dots i_{b-1}i_b i_{b+1}\dots i_p}=-A_{i_1\dots i_{a-1}i_b i_{a+1}\dots i_{b-1}i_a i_{b+1}\dots i_p}.$$

The coefficient of $$\frac{\partial(x_{i_1},\dots,x_{i_p})}{\partial(u_1,\dots,u_p)}$$

in the integral in $(u_1,\dots,u_p)$ which defines the value of our integral over E_p is therefore $p!\,A_{i_1\dots i_p}$, and it is convenient, in order to avoid complications arising from numerical coefficients, to write all multiple integrals in the form

$$\int \frac{1}{p!} A_{i_1 \dots i_p} dx^{i_1} \dots dx^{i_p}, \tag{1}$$

where the coefficients are skew-symmetric. This convention will be adhered to throughout.

19·3. The definition of a p-fold integral in a simplicial region D refers to a particular coordinate system $(x_1, \dots, x_n)$ in D. Let $(x'_1, \dots, x'_n)$ be any other coordinate system valid in D, obtained from $(x_1, \dots, x_n)$ by a transformation of class *one* at least. Let

$$\int \frac{1}{p!} A'_{i_1 \dots i_p} dx'^{i_1} \dots dx'^{i_p} \tag{2}$$

be a p-fold integral defined in D, relative to the coordinate system $(x'_1, \dots, x'_n)$. We seek necessary and sufficient conditions that the value of this integral over every p-simplex E_p of D should be equal to the value of (1) over E_p. The necessary and sufficient condition is that

$$\bar{A}'_{i_1 \dots i_p} \frac{\partial(x'_{i_1}, \dots, x'_{i_p})}{\partial(u_1, \dots, u_p)} = \bar{A}_{i_1 \dots i_p} \frac{\partial(x_{i_1}, \dots, x_{i_p})}{\partial(u_1, \dots, u_p)},$$

for all substitutions

$$x_i = x_i(u_1, \dots, u_p), \quad x'_i = x'_i\{x_1(u), \dots, x_n(u)\} = x'_i(u_1, \dots, u_p)$$
$$(i = 1, \dots, n),$$

where

$$\bar{A}_{i_1 \dots i_p} = A_{i_1 \dots i_p}\{x_1(u), \dots, x_n(u)\},$$
$$\bar{A}'_{i_1 \dots i_p} = A'_{i_1 \dots i_p}\{x'_1(u), \dots, x'_n(u)\}.$$

In particular, if E_p satisfies

$$x'_{j_s} = \text{constant} \qquad (s = p+1, \dots, n),$$

we may take

$$u_r = x'_{j_r} \qquad (r = 1, \dots, p).$$

We then have the conditions

$$p!\, A'_{j_1 \dots j_p} = A_{i_1 \dots i_p} \frac{\partial(x_{i_1}, \dots, x_{i_p})}{\partial(x'_{j_1}, \dots, x'_{j_p})},$$

which may be written in the form

$$A'_{j_1 \dots j_p} = A_{i_1 \dots i_p} \frac{\partial x_{i_1}}{\partial x'_{j_1}} \cdots \frac{\partial x_{i_p}}{\partial x'_{j_p}},$$

since $A_{i_1 \dots i_p}$ is skew-symmetric in the suffixes. This relation must hold for all $(j_1, \dots, j_p)$.

Conversely, suppose this relation holds for all $(j_1, \dots, j_p)$. Then, for all substitutions,

$$x_i = x_i(u_1, \dots, u_p); \quad x'_i = x'_i\{x_1(u), \dots, x_n(u)\} = x'_i(u_1, \dots, u_p)$$
$$(i = 1, \dots, n),$$

in which the functions $x_i(u_1, \dots, u_p)$ are differentiable, we have

$$\begin{aligned}
\frac{1}{p!}\bar{A}'_{i_1 \dots i_p} \frac{\partial(x'_{i_1}, \dots, x'_{i_p})}{\partial(u_1, \dots, u_p)} &= \bar{A}'_{i_1 \dots i_p} \frac{\partial x'_{i_1}}{\partial u_1} \cdots \frac{\partial x'_{i_p}}{\partial u_p} \\
&= \bar{A}_{j_1 \dots j_p} \frac{\partial x_{j_1}}{\partial x'_{i_1}} \cdots \frac{\partial x_{j_p}}{\partial x'_{i_p}} \frac{\partial x'_{i_1}}{\partial u_1} \cdots \frac{\partial x'_{i_p}}{\partial u_p} \\
&= \bar{A}_{j_1 \dots j_p} \frac{\partial x_{j_1}}{\partial u_1} \cdots \frac{\partial x_{j_p}}{\partial u_p} \\
&= \frac{1}{p!}\bar{A}_{j_1 \dots j_p} \frac{\partial(x_{j_1}, \dots, x_{j_p})}{\partial(u_1, \dots, u_p)},
\end{aligned}$$

and it follows that the integrals (1) and (2) have the same values over every p-simplex E_p in D.

The relation connecting the coefficients $A_{i_1 \dots i_p}$, $A'_{j_1 \dots j_p}$ is one already discussed in connection with the theory of tensors. We therefore conclude that in order to define a p-fold integral in a simplicial domain D, of class *one* at least, we require an absolute skew-symmetric covariant tensor field of rank p, where the components are continuous functions in D. The tensor field defines, in each coordinate system, a p-fold integral in D, and the set of integrals so obtained, one in each coordinate system, has the property that all integrals of the set have the same values over any p-simplex E_p of D. We say that the integrals of the set are equivalent, and that the different integrals of the set give the representations, in the different coordinate systems, of the same integral in D.

19·4. We can now define an integral in any region R of a manifold M, where R need not be simplicial, and may be the whole manifold. The region R can be covered by a finite number of neighbourhoods of M, say, $N_1, \ldots, N_r$. If P is any point of R which lies in $N_{i_1}, \ldots, N_{i_s}$, but not in the remaining N_j, there is a neighbourhood of P which lies in $N_{i_1}, \ldots, N_{i_s}, R$. We denote such a neighbourhood by $N_R(P)$. An integral defined in $N_{i_j} (j \leqslant s)$ defines an integral in $N_R(P)$. Suppose, now, that we have a series of integrals, one defined in each N_i. If the integrals in $N_{i_1}, \ldots, N_{i_s}$ all define the same integral in $N_R(P)$, for every point P of R, we say that the integrals in $N_{i_1}, \ldots, N_{i_s}$ define an integral in R. It is easily verified that this definition does not depend on the set of neighbourhoods of M chosen to cover R.

A skew-symmetric covariant tensor of rank p in a space of n dimensions is always zero if p is greater than n. Hence, on an n-dimensional manifold we have to consider integrals of multiplicity p, where $1 \leqslant p \leqslant n$. It is convenient to introduce, in addition, integrals of multiplicity zero. An 0-fold integral is defined by means of a single function on the manifold. An 0-simplex is simply a point, and we define the value of the integral $\int f(x)$ over the 0-simplex P by the equality

$$\int_P f(x) = f(P).$$

In our investigations of the properties of integrals on a manifold, it is convenient to regard the functions which appear as functions of the points of the manifold rather than of the local coordinates, so that the same function is denoted by the same symbol whatever coordinate system is used. A transformation of a tensor field is then a transformation of one set of functions of the points of a manifold into another set of functions of the points of the manifold.

20·1. **The theorem of Stokes.** In order to define a multiple integral on a manifold of class u, we require an absolute skew-symmetric covariant tensor field, whose components are continuous functions of the local parameters. In the remainder of this chapter it will be necessary to assume that the components of the tensor field possess a certain number of continuous derivatives. We shall not always state the class v of the components explicitly, but it is to be understood that v ($\leqslant u$) is sufficiently high for our results to have a meaning. If we say that an integral is *regular* in a region R, we imply that the integral is defined at all points of R, and that its coefficients are of class *two* at least at every point of R.

We have so far defined the value of an integral only for oriented simplexes. It is easy to pass to the definition of the value of a p-fold integral on a manifold over any p-chain of the manifold. Let

$$\int \frac{1}{p!} A_{i_1 \ldots i_p} dx^{i_1} \ldots dx^{i_p}$$

be a p-fold integral on a manifold M, and let

$$C_p = a_i E^i_p$$

be a p-chain on M, of class *one* at least. We define the value of the integral over C_p by the equality

$$\int_{C_p} \frac{1}{p!} A_{i_1 \ldots i_p} dx^{i_1} \ldots dx^{i_p} = \sum_i a_i \int_{E^i_p} \frac{1}{p!} A_{i_1 \ldots i_p} dx^{i_1} \ldots dx^{i_p}.$$

From the definition and properties of a multiple integral given by Goursat, we obtain the property that the value of a multiple integral taken over a p-chain is unaltered by subdivision of the chain.

20·2. The theorem of Green has been proved by Goursat for simple integrals in space of two dimensions, and for double integrals in space of three dimensions. The same methods enable us to prove it for $(p-1)$-fold integrals in a number space of p dimensions; we obtain the result that

$$\int \sum_{i=1}^{p} (-1)^{i-1} P_i \, du^1 \ldots du^{i-1} du^{i+1} \ldots du^p,$$

evaluated over the exterior boundary of the domain $\mathscr{D}$, is equal to

$$\int_{\mathscr{D}} \sum_{i=1}^{p} \frac{\partial P_i}{\partial u_i} du^1 \dots du^p.$$

Take $\mathscr{D}$ to be a simplex $\mathscr{E}_p$, and orient it so that the coordinate system $(u_1, \dots, u_p)$ is concordant with the orientation. If we denote the oriented cell by E_p and its oriented boundary by $F(E_p)$, the theorem can be written

$$\int_{F(E_p)} \sum_{i=1}^{p} (-1)^{i-1} P_i \, du^1 \dots du^{i-1} du^{i+1} \dots du^p$$
$$= \int_{E_p} \sum_{i=1}^{p} \frac{\partial P_i}{\partial u_i} du^1 \dots du^p.$$

Since
$$F(-E_p) = -F(E_p),$$
this result is true whether the coordinate system is concordant with the orientation of the cell or not.

Now consider a general simplex E_p of our n-space, given by

$$x_i = x_i(u_1, \dots, u_p) \qquad (i = 1, \dots, n),$$

where $(u_1, \dots, u_p)$ is a parametric system concordant with the orientation, and $(u_1, \dots, u_p)$ vary in a simplex E'_p, which we suppose oriented by these parameters. Then consider the integral

$$\int \frac{1}{(p-1)!} A_{i_1 \dots i_{p-1}} dx^{i_1} \dots dx^{i_{p-1}},$$

evaluated over the boundary $F(E_p)$ of E_p. It is equal to

$$\int_{F(E'_p)} \frac{1}{(p-1)!} \sum_{j=1}^{p} A_{i_1 \dots i_{p-1}} \frac{\partial(x_{i_1}, \dots, x_{i_{p-1}})}{\partial(u_1, \dots, u_{j-1}, u_{j+1}, \dots, u_p)}$$
$$\times du^1 \dots du^{j-1} du^{j+1} \dots du^p$$
$$= \int_{E'_p} \frac{1}{(p-1)!} \sum_{j=1}^{p} (-1)^{j-1}$$
$$\times \frac{\partial}{\partial u_j} \left[A_{i_1 \dots i_{p-1}} \frac{\partial(x_{i_1}, \dots, x_{i_{p-1}})}{\partial(u_1, \dots, u_{j-1}, u_{j+1}, \dots, u_p)} \right] du^1 \dots du^p$$
$$= \int_{E'_p} \frac{1}{(p-1)!} \sum_{k=1}^{n} \frac{\partial}{\partial x_k} A_{i_1 \dots i_{p-1}} \frac{\partial(x_k, x_{i_1}, \dots, x_{i_{p-1}})}{\partial(u_1, \dots, u_p)} du^1 \dots du^p,$$

since, by a familiar theorem,

$$\sum_{j=1}^{p}(-1)^{j-1}\frac{\partial}{\partial u_j}\frac{\partial(x_{i_1},\ldots,x_{i_{p-1}})}{\partial(u_1,\ldots,u_{j-1},u_{j+1},\ldots,u_p)}=0.$$

If we write

$$B_{i_1\ldots i_p}=\sum_{k=1}^{p}(-1)^{k-1}\frac{\partial}{\partial x_{i_k}}A_{i_1\ldots i_{k-1}i_{k+1}\ldots i_p},$$

$$\int_{E'_p}\frac{1}{p!}B_{i_1\ldots i_p}\frac{\partial(x_{i_1},\ldots,x_{i_p})}{\partial(u_1,\ldots,u_p)}du^1\ldots du^p$$

$$=\int_{E'_p}\frac{1}{(p-1)!}\sum_{k=1}^{n}\frac{\partial}{\partial x_k}A_{i_1\ldots i_{p-1}}\frac{\partial(x_k,x_{i_1},\ldots,x_{i_{p-1}})}{\partial(u_1,\ldots,u_p)}du^1\ldots du^p,$$

and hence

$$\int_{E_p}\frac{1}{p!}B_{i_1\ldots i_p}dx^{i_1}\ldots dx^{i_p}=\int_{F(E_p)}\frac{1}{(p-1)!}A_{i_1\ldots i_{p-1}}dx^{i_1}\ldots dx^{i_{p-1}}.$$

From the form of this result it is clear that it does not depend on the particular simplex E_p considered, or on the particular rectilinear cell E'_p associated with it. Moreover, the result extends at once to any chain, and we have

$$\int_{C_p}\frac{1}{p!}B_{i_1\ldots i_p}dx^{i_1}\ldots dx^{i_p}=\int_{F(C_p)}\frac{1}{(p-1)!}A_{i_1\ldots i_{p-1}}dx^{i_1}\ldots dx^{i_{p-1}}.$$

This theorem is the *Theorem of Stokes*, in its most general form; it holds for $1\leqslant p\leqslant n$. If we make a change of coordinate system, writing

$$x_i=x_i(\bar{x}_1,\ldots,\bar{x}_n)\qquad(i=1,\ldots,n),$$

the integral $\displaystyle\int\frac{1}{(p-1)!}A_{i_1\ldots i_{p-1}}dx^{i_1}\ldots dx^{i_{p-1}}$

becomes $\displaystyle\int\frac{1}{(p-1)!}\bar{A}_{i_1\ldots i_{p-1}}d\bar{x}^{i_1}\ldots d\bar{x}^{i_{p-1}},$

and $\displaystyle\sum_{r=1}^{p}(-1)^{r-1}\frac{\partial}{\partial\bar{x}_{i_r}}\bar{A}_{i_1\ldots i_{r-1}i_{r+1}\ldots i_p}$

$$=\sum_{r=1}^{p}(-1)^{r-1}\frac{\partial}{\partial\bar{x}_{i_r}}\left[A_{j_1\ldots j_{p-1}}\frac{\partial x_{j_1}}{\partial\bar{x}_{i_1}}\cdots\frac{\partial x_{j_{r-1}}}{\partial\bar{x}_{i_{r-1}}}\frac{\partial x_{j_r}}{\partial\bar{x}_{i_{r+1}}}\cdots\frac{\partial x_{j_{p-1}}}{\partial\bar{x}_{i_p}}\right]$$

$$=\left[\sum_{k=1}^{p}(-1)^{k-1}\frac{\partial}{\partial x_{j_k}}A_{j_1\ldots j_{k-1}j_{k+1}\ldots j_p}\right]\frac{\partial x_{j_1}}{\partial\bar{x}_{i_1}}\cdots\frac{\partial x_{j_p}}{\partial\bar{x}_{i_p}},$$

since the terms involving second derivatives cancel.

This verifies that the relation between the $(p-1)$-fold integral and the p-fold integral is invariant under change of coordinate system, a result which is obvious geometrically.

21. **Calculus of forms.** In order to express properties of integrals, and relations between integrals, in a convenient manner, it is desirable to develop a calculus of the expressions which appear as the integrands of our integrals. Following Cartan, we call an expression such as

$$\frac{1}{p!}A_{i_1\ldots i_p}dx^{i_1}\ldots dx^{i_p}$$

a *p-form*, and denote it by A, or, if we wish to stress its multiplicity p, or the set of variables in which it is expressed, or both, we denote it by A_p, or $A(x)$, or $A_p(x)$. Different forms are denoted by different letters $A, B, \ldots$, or by indices A_p^i, etc.

If
$$A = \frac{1}{p!}A_{i_1\ldots i_p}dx^{i_1}\ldots dx^{i_p}$$

is any p-form, the $(p+1)$-form obtained from it by Stokes' theorem is denoted by A_x and we call it the *derived form*, or the *exterior derivative*, of A. Thus Stokes' theorem is expressed by the equation

$$\int_{F(C_{p+1})} A = \int_{C_{p+1}} A_x.$$

We now develop a calculus of forms.

I. Two forms of the same multiplicity can be added.

If
$$A = \frac{1}{p!}A_{i_1\ldots i_p}dx^{i_1}\ldots dx^{i_p}$$

and
$$B = \frac{1}{p!}B_{i_1\ldots i_p}dx^{i_1}\ldots dx^{i_p},$$

then
$$A+B = \frac{1}{p!}[A_{i_1\ldots i_p}+B_{i_1\ldots i_p}]dx^{i_1}\ldots dx^{i_p}.$$

II. We define the product $A_p \times B_q$ of two forms

$$A_p = \frac{1}{p!} A_{i_1 \ldots i_p} dx^{i_1} \ldots dx^{i_p}$$

and
$$B_q = \frac{1}{q!} B_{i_1 \ldots i_q} dx^{i_1} \ldots dx^{i_q}$$

as follows:

(i) if $p+q>n$, $\quad A_p \times B_q = 0;$

(ii) if $p+q \leqslant n$, $\quad A_p \times B_q$ is the $(p+q)$-form

$$\frac{1}{(p+q)!} C_{i_1 \ldots i_{p+q}} dx^{i_1} \ldots dx^{i_{p+q}},$$

where
$$C_{i_1 \ldots i_{p+q}} = \frac{1}{p!}\frac{1}{q!} \delta^{j_1 \ldots j_p k_1 \ldots k_q}_{i_1 \ldots \ldots \ldots \ldots i_{p+q}} A_{j_1 \ldots j_p} B_{k_1 \ldots k_q}.$$

From these definitions the following results come immediately:

(i) $A_p + (B_p + C_p) = (A_p + B_p) + C_p = A_p + B_p + C_p.$

(ii) $A_p + B_p = B_p + A_p.$

(iii) $A_p \times (B_q + C_q) = A_p \times B_q + A_p \times C_q.$

(iv) $A_p \times (B_q \times C_r) = (A_p \times B_q) \times C_r = A_p \times B_q \times C_r.$

(v) $A_p \times B_q = (-1)^{pq} B_q \times A_p.$

(vi) $(A_p \times B_q)_x = (A_p)_x \times B_q + (-1)^p A_p \times (B_q)_x.$

(vii) $(A_x)_x = 0.$

The first five of these identities require no restrictions on the coefficients of the forms, the sixth requires that the coefficients be of class *one*, and the last requires the coefficients to be of class *two*.

22·1. **Periods.** So far, we have only used certain local properties of the manifold on which the integrals are given. We now consider any absolute manifold M of class u, which is orientable, i.e. a Riemannian manifold of class u. We do not at present make any use of the Riemannian metric, but, as e pointed out in § 3, we can always impose on a manifold of class u, which satisfies the other properties of a Riemannian manifold, a Riemannian metric of a *finite* class $v \leqslant u-1$, so no greater degree of generality is achieved by omitting the metric.

We now wish to consider properties of integrals in relation to the manifold as a whole. We shall assume that the forms which we consider are regular.

The theorem of Stokes expresses the integral of a p-form A over a bounding-cycle Γ as the integral of the derived form A_x over a chain whose boundary is Γ. From it we deduce that if the derived form is identically zero, the value of $\int A$ over any bounding-cycle is zero. Now consider a fundamental base

$$\Gamma_p^i \qquad (i = 1, \ldots, R_p)$$

for the p-cycles of the manifold M. Let Γ_p be any p-cycle of M, satisfying the homology

$$\Gamma_p \approx \lambda_i \Gamma_p^i,$$

where the coefficients λ_i are integers. Then, for a suitable integer $\lambda \neq 0$, there exists a $(p+1)$-chain C_{p+1} such that

$$C_{p+1} \rightarrow \lambda \Gamma_p - \lambda \lambda_i \Gamma_p^i.$$

Hence, if
$$A_x = 0,$$
we have
$$\int_{\lambda \Gamma_p} A = \int_{\lambda \lambda_i \Gamma_p^i} A,$$
that is,
$$\int_{\Gamma_p} A = \sum_{i=1}^{R_p} \lambda_i \int_{\Gamma_p^i} A.$$
Let us write
$$\int_{\Gamma^i} A = \omega^i.$$

Let γ_{p-1} be any bounding-cycle of $(p-1)$ dimensions of M. The p-chain which it bounds is not uniquely defined, but if C_p and C'_p are two chains having γ_{p-1} as boundary, we have

$$C'_p - C_p \rightarrow 0,$$

that is, $C'_p - C_p$ is a cycle, given, say, by the homology

$$C'_p - C_p \approx \lambda_i \Gamma^i_p.$$

Hence
$$\int_{C'_p} A = \int_{C_p} A + \lambda_i \omega^i.$$

It follows that the integral of a p-form whose exterior derivative is identically zero defines a "functional" of the bounding $(p-1)$-cycles of M, unique save for a linear function

$$\lambda_i \omega^i$$

of certain constants ω^i, with integral coefficients λ_i. These constants are *periods* of the integral *on*, or *with respect to*, the cycles Γ^i_p. If Γ_p is any p-cycle such that

$$\Gamma_p \approx \lambda_i \Gamma^i_p,$$

then the period of the integral on the cycle Γ_p is

$$\int_{\Gamma_p} A = \lambda_i \omega^i.$$

The period is always zero on a bounding-cycle or a divisor of zero.

22·2. Forms whose exterior derivatives are zero will be called *closed forms*, and in place of

$$A_x = 0,$$

we shall usually write
$$A \rightarrow 0.$$

A closed form of multiplicity zero is therefore a constant, and any form of multiplicity n is closed. It is to be remembered that it is only in connection with closed forms that we can speak of periods.

A special case of a closed form suggests itself at once. Let B be any regular $(p-1)$-form and let

$$A = B_x.$$

Then
$$A \rightarrow 0.$$

For such forms, the periods of the integrals are all zero. For, let us consider any p-chain C_p. Then

$$\int_{C_p} A = \int_{F(C_p)} B.$$

Hence, if C_p is a cycle, $\int_{C_p} A = 0,$

and so $\int_{\Gamma_p^i} A = 0 \qquad (i = 1, \ldots, R_p).$

Forms which are the exterior derivatives of other forms are called *null forms*. If A is a null form, we write

$$A \sim 0,$$

and if A and B are two forms of the same multiplicity which differ by a null form, we write

$$A \sim B$$

and speak of them as *homologous* forms. Homologous closed forms have the same periods and null forms have all their periods zero. We shall see later that the converse of these results holds, that is, that if two closed forms have the same periods they are homologous, and that if a closed form has all its periods zero it is null.

22·3. Let M' be a manifold homeomorphic with M, and denote any object on M' by the same symbol as the corresponding object on M, with, however, a "prime" attached. Construct the product manifold $M \times M'$. The topology of this has been discussed in § 18, and we adopt here the same notation as in that paragraph.

Let A_p be any p-form, and B_q a q-form, both regular on M. We can construct from these certain forms on $M \times M'$. At any point $P \times Q'$ of $M \times M'$ we can obtain a coordinate system $(x_1, \ldots, x_n; y'_1, \ldots, y'_n)$ by combining the coordinate system $(x_1, \ldots, x_n)$ at P on M with the coordinate system $(y'_1, \ldots, y'_n)$ at Q' on M'. We can define a p-form on $M \times M'$ by the condition that in the coordinate system $(x_1, \ldots, x_n; y'_1, \ldots, y'_n)$ at $P \times Q'$ it is represented by

$$\frac{1}{p!} A_{i_1 \ldots i_p} dx^{i_1} \ldots dx^{i_p},$$

where this expression also represents the form A_p on M in the coordinate system $(x_1, \ldots, x_n)$.

We denote this form on $M \times M'$ by A_p. Similarly, from the form B'_q on M' we define a form B'_q on $M \times M'$. The product of these forms is a $(p+q)$-form $A_p \times B'_q$ on $M \times M'$.

We first show how to evaluate the integral of $A_p \times B'_q$ over a $(p+q)$-cell on $M \times M'$ which is the product of an r-simplex E_r of M by an s-simplex E'_s of M', where $p+q = r+s$. We suppose that E_r, E'_s are so small that we can find coordinates $(x_1, \ldots, x_n)$ on M and $(y'_1, \ldots, y'_n)$ on M', such that E_r lies in the locus

$$x_{r+1} = 0, \ldots, x_n = 0,$$

and E'_s lies in $\quad y'_{s+1} = 0, \ldots, y'_n = 0.$

In these coordinates let

$$A_p = \frac{1}{p!} A_{i_1 \ldots i_p} dx^{i_1} \ldots dx^{i_p},$$

$$B'_q = \frac{1}{q!} B'_{i_1 \ldots i_q} dy'^{i_1} \ldots dy'^{i_q}.$$

In the coordinate system $(x_1, \ldots, x_n; y'_1, \ldots, y'_n)$ on $M \times M'$ we have

$$A_p \times B_q = \frac{1}{p!} \frac{1}{q!} A_{i_1 \ldots i_p} B'_{j_1 \ldots j_q} dx^{i_1} \ldots dx^{i_p} dy'^{j_1} \ldots dy'^{j_q},$$

the summations being from 1 to n. To evaluate the integral of this form over $E_r \times E'_s$ we put

$$dx^{r+1} = \ldots = dx^n = 0, \quad dy'^{s+1} = \ldots = dy'^n = 0.$$

If $p \neq r$, $q \neq s$, this reduces $A_p \times B'_q$ to zero. Hence

$$\int_{E_r \times E'_s} A_p \times B'_q = 0.$$

On the other hand if $p = r$, $q = s$, we have

$$\begin{aligned}\int_{E_p\times E'_q} A_p \times B'_q &= \int_{E_p\times E'_q} A_{1\ldots p} B'_{1\ldots q}\, dx^1 \ldots dx^p\, dy'^1 \ldots dy'^q \\ &= \int_{E_p} A_{1\ldots p}\, dx^1 \ldots dx^p \times \int_{E'_q} B'_{1\ldots q}\, dy'^1 \ldots dy'^q \\ &= \int_{E_p} A_p \times \int_{E'_q} B'_q \\ &= \int_{E_p} A_p \times \int_{E_q} B_q.\end{aligned}$$

From the definition of the product of two chains we deduce immediately that if $p+q = r+s$,

$$\int_{C_r\times C'_s} A_p \times B'_q = 0 \qquad (p\neq r, q\neq s),$$

and

$$\begin{aligned}\int_{C_p\times C'_q} A_p \times B'_q &= \int_{C_p} A_p \times \int_{C'_q} B'_q \\ &= \int_{C_p} A_p \times \int_{C_q} B_q.\end{aligned}$$

The most important case of the foregoing result occurs when A_p and B_q are both closed. Then

$$(A_p \times B'_q)_x = (A_p)_x \times B'_q + (-1)^p A_p \times (B'_q)_{y'} = 0.$$

The product form is thus closed, and we can speak of its periods. If Γ_r, Γ_s are two cycles of M, where $p+q = r+s$, $\Gamma_r \times \Gamma'_s$ is a cycle of $M\times M'$, and we have

$$\int_{\Gamma_r\times \Gamma'_s} A_p \times B'_q = 0 \qquad (p\neq r, q\neq s),$$

and

$$\int_{\Gamma_p\times \Gamma'_q} A_p \times B'_q = \int_{\Gamma_p} A_p \times \int_{\Gamma_q} B_q.$$

22·4. We now consider the case in which A_p and B_q are closed, and $p+q = n$. Let

$$\Gamma^i_r \qquad (i = 1, \ldots, R_r)$$

be a fundamental base for the r-cycles of M, for $r=0, \ldots, n$. A base for the n-cycles of $M \times M'$ is given by

$$\Gamma_r^i \times \Gamma'^j_{n-r} \qquad (i=1, \ldots, R_r;\ j=1, \ldots, R_{n-r};\ r=0, \ldots, n).$$

We denote the periods of the integral of A_p on the cycles Γ_p^i of the base for the p-cycles of M by ω^i,

$$\omega^i = \int_{\Gamma_p^i} A_p,$$

and we denote the periods of the integral of B_{n-p} on the cycles Γ_{n-p}^i by ν^i. The periods of

$$\int A_p \times B'_{n-p}$$

on the base for the n-cycles of $M \times M'$ are therefore known at once by our formulae.

Let Γ be the diagonal cycle on $M \times M'$ given (§ 18·3) by

$$\Gamma \approx \sum_{r=0}^{n} \epsilon_{ij}^r \Gamma_r^i \times \Gamma'^j_{n-r}.$$

Then

$$\int_\Gamma A_p \times B'_{n-p} = \sum_{r,i,j} \epsilon_{ij}^r \int_{\Gamma_r^i \times \Gamma'^j_{n-r}} A_p \times B'_{n-p} = \epsilon_{ij}^p \omega^i \nu^j.$$

Now any point of P is expressible in the form $P \times P'$, in the notation which we are using, and we can choose the local coordinate systems $(x_1, \ldots, x_n)$ and $(x_1', \ldots, x_n')$ on M and M' so that Γ is given in the coordinate system $(x_1, \ldots, x_n;\ x_1', \ldots, x_n')$ on $M \times M'$ by the equations

$$x_i = x_i' \qquad (i=1, \ldots, n).$$

It follows that

$$\int_\Gamma A_p \times B'_{n-p} = \int_M A_p \times B_{n-p}$$

and we obtain the formula

$$\int_M A_p \times B_{n-p} = \epsilon_{ij}^p \omega^i \nu^j.$$

The matrix $\boldsymbol{\epsilon}^p$ is the transpose of the inverse of the intersection matrix $\|(\Gamma_p^i \,.\, \Gamma_{n-p}^j)\|$.

The applications of this result are of frequent occurrence in the sequel, and we shall refer to the above equation as the *bilinear relation* connecting the periods of the two integrals. If we have to consider a number of closed p-forms

$$A_p^1, \ldots, A_p^s$$

and a number of closed $(n-p)$ forms

$$B_{n-p}^1, \ldots, B_{n-p}^t,$$

we shall find it convenient to write the st bilinear relations in the following matrix form. If

$$\omega^{ij} = \int_{\Gamma_p^j} A_p^i,$$

and

$$\nu^{ij} = \int_{\Gamma_{n-p}^j} B_{n-p}^i,$$

the matrices $\boldsymbol{\omega}$ and $\boldsymbol{\nu}$ are called period matrices. The st bilinear relations are all contained in the matrix equation

$$\left\| \int_M A_p^i \times B_{n-p}^j \right\| = \boldsymbol{\omega}\boldsymbol{\epsilon}^p\boldsymbol{\nu}',$$

$\boldsymbol{\nu}'$ indicating the transpose of the matrix $\boldsymbol{\nu}$.

In place of the fundamental bases

$$\Gamma_p^i \quad (i=1, \ldots, R_p); \qquad \Gamma_{n-p}^j \quad (j=1, \ldots, R_{n-p})$$

for the p-cycles and $(n-p)$-cycles we can consider new bases, not necessarily fundamental, given by

$$\Delta_p^i = a_j^i \Gamma_p^j \qquad (i=1, \ldots, R_p)$$

and

$$\Delta_{n-p}^i = b_j^i \Gamma_{n-p}^j \qquad (i=1, \ldots, R_{n-p}),$$

where $\mathbf{a}$, $\mathbf{b}$ are non-singular matrices of integers. The period matrices of the integrals for the new bases are

$$\boldsymbol{\omega}_1 = \boldsymbol{\omega}\mathbf{a}', \quad \boldsymbol{\nu}_1 = \boldsymbol{\nu}\mathbf{b}',$$

and the bilinear equations can be written as

$$\left\| \int_M A_p^i \times B_{n-p}^j \right\| = \boldsymbol{\omega}_1 \boldsymbol{\theta}^p \boldsymbol{\nu}_1',$$

where $$\mathbf{a}'\boldsymbol{\theta}^p\mathbf{b} = \boldsymbol{\epsilon}^p.$$

Now the intersection matrix of the new bases is

$$\mathbf{a} \| (\Gamma_p^i . \Gamma_{n-p}^j) \| \mathbf{b}',$$

and hence $\boldsymbol{\theta}^p$ is the transpose of the inverse of the intersection matrix $\| (\Delta_p^i . \Delta_{n-p}^j) \|$ of the cycles Δ_p^i and Δ_{n-p}^j. Thus the bilinear relations hold whatever base we take for the period cycles, fundamental or not.

22·5. More generally, let A_p, B_q be two closed forms on M, such that $p+q = r \leqslant n$. Then, as before, we show that

$$\int_{\Gamma_r^h} A_p \times B_q = \int_{\gamma_r^h} A_p \times B_q',$$

where γ_r^h is the image of Γ_r^h on the diagonal cycle of $M \times M'$. If we refer to § 18·3 we see that

$$\gamma_r^h \approx \sum_s \lambda_{ij}^s (\Gamma_r^h) \Gamma_{r+s-n}^i \times \Gamma_{n-s}'^j.$$

Hence $$\int_{\Gamma_r^h} A_p \times B_q = \sum_{i,j} \lambda_{ij}^{(n-q)} (\Gamma_r^h) \int_{\Gamma_p^i \times \Gamma_q'^j} A_p \times B_q'$$

$$= \sum_{i,j} \lambda_{ij}^{(n-q)} (\Gamma_r^h) \int_{\Gamma_p^i} A_p \times \int_{\Gamma_q^j} B_q.$$

The values of the coefficients $\lambda_{ij}^{(n-q)} (\Gamma_r^h)$ are given by formula (12) of § 18·3.

We can extend this method to find the periods of the product of three or more closed forms.

23·1. **The first theorem of de Rham.** In discussing the period properties of the integrals of closed forms, we have tacitly assumed the existence of a closed p-form whose integral has a period different from zero on some p-cycle of our manifold. The justification of this assumption is contained in the first of two theorems due to de Rham [2], we shall refer to these theorems as *de Rham's first theorem* and *de Rham's second theorem*, respectively. At the moment we are concerned only with *de Rham's first theorem*, the statement of which is as follows:

If Γ_p^i $(i=1, \dots, R_p)$ is any base for the p-cycles of a manifold M, and ν^i $(i=1, \dots, R_p)$ are R_p arbitrary real numbers, there exists a p-form ϕ with the properties:

(i) *ϕ is regular and closed on M; and*

(ii) $$\int_{\Gamma_p^i} \phi = \nu^i \qquad (i=1, \dots, R_p).$$

The proof of the theorem which we give is essentially that given by de Rham, but for the sake of clarity we begin with some preliminary considerations which deal with an algebraic approximation to de Rham's theorem.

23·2. *p-sets.* Let K be a complex covering our manifold M. We denote its p-simplexes by E_p^i $(i=1, \dots, R_p)$, and its incidence matrices by ${}_{(p)}\eta$, as usual. If ϕ is any p-form on M, the value of the integral of ϕ over E_p^i is a real number, which we denote by e_p^i. We shall say that the integral is represented by the set of numbers $(e_p^1, \dots, e_p^{\alpha_p})$. Any set $(e_p^1, \dots, e_p^{\alpha_p})$, where e_p^i is a real number, is called a *p-set*.

The $(p+1)$-set $(e_{p+1}^1, \dots, e_{p+1}^{\alpha_{p+1}})$ which represents the integral of the derived form ϕ_x can be determined at once. For,

$$\begin{aligned} e_{p+1}^i &= \int_{E_{p+1}^i} \phi_x \\ &= \int_{F(E_{p+1}^i)} \phi \\ &= {}_{(p+1)}\eta_j^i \, e_p^j. \end{aligned}$$

This $(p+1)$-set $(e_{p+1}^1, \dots, e_{p+1}^{\alpha_{p+1}})$ is called the derived set of the set $(e_p^1, \dots, e_p^{\alpha_p})$. If $a_i E_{p+1}^i$ is any $(p+1)$-cycle, we have

$$\begin{aligned} a_i e_{p+1}^i &= a_i \, {}_{(p+1)}\eta_j^i \, e_p^j \\ &= 0. \end{aligned}$$

We now show that, conversely, if $(e^1_{p+1}, \ldots, e^{\alpha_{p+1}}_{p+1})$ is a $(p+1)$-set such that

$$a_i e^i_{p+1} = 0$$

for all sets of integers a_i for which $a_i E^i_{p+1}$ is a $(p+1)$-cycle, then this $(p+1)$-set is a derived set. In order to do this we have to solve the equations

$$e^i_{p+1} = {}_{(p+1)}\eta^i_j e^j_p \tag{3}$$

for the unknowns e^i_p. The necessary and sufficient condition that the equations (3) have a solution is that

$$b_i e^i_{p+1} = 0$$

for all numbers b_i satisfying

$$b_{i\,(p+1)}\eta^i_j = 0 \qquad (j = 1, \ldots, \alpha_p). \tag{4}$$

Since the matrix ${}_{(p+1)}\eta$ is a matrix of integers, we need only consider integral values of b_i. Now (4) is the condition to be satisfied in order that $b_i E^i_{p+1}$ should be a $(p+1)$-cycle. From our hypothesis it follows that the conditions necessary for the existence of a solution of (3) are satisfied. The result follows at once.

A derived p-set has the further property that its derived $(p+1)$-set is zero. For, if

$$e^i_p = {}_{(p)}\eta^i_j e^j_{p-1},$$

then

$$\begin{aligned} {}_{(p+1)}\eta^i_j e^j_p &= {}_{(p+1)}\eta^i_j\, {}_{(p)}\eta^j_k e^k_{p-1} \\ &= 0, \end{aligned}$$

by equation (7), § 12·2 of Chapter I.

The converse result, that a p-set whose derived set is zero is necessarily a derived set, is not generally true. A p-set whose derived $(p+1)$-set is zero is called a *closed* p-set.

23·3. We now consider certain properties of a closed p-set $(e^1_p, \ldots, e^{\alpha_p}_p)$. If $a_i E^i_p$ is any p-cycle which is homologous to zero with division, there exist integers $\lambda, b_1, \ldots, b_{\alpha_{p+1}}$, such that

$$b_{i\,(p+1)}\eta^i_j = \lambda a_j \qquad (\lambda \neq 0).$$

Then

$$a_j e^j_p = \lambda^{-1} b_{i\,(p+1)}\eta^i_j e^j_p$$
$$= 0.$$

It follows that if $a_i E^i_p$, $b_i E^i_p$ are two p-cycles of K homologous to each other with division, we have

$$(a_i - b_i)\, e^i_p = 0,$$

that is,

$$a_i e^i_p = b_i e^i_p.$$

We call the number $a_i e^i_p$ the *period* of the p-set over the cycle $a_i E^i_p$, and our result means that the periods of a closed p-set over p-cycles of K which are homologous with division are equal.

A derived p-set is a closed set, and the result proved in § 23·2 shows that the necessary and sufficient conditions that a p-set should be a derived set are that it should be closed and have all its periods zero.

The theorem for p-sets analogous to de Rham's theorem for integrals states that if Γ^i_p $(i = 1, \ldots, R_p)$ is a base for the p-cycles of K, and ν^i_p $(i = 1, \ldots, R_p)$ are R_p arbitrary real numbers, there exists a p-set $(e^1_p, \ldots, e^{\alpha_p}_p)$ which is closed and has the period ν^i on Γ^i_p $(i = 1, \ldots, R_p)$. We now prove this result. The equations which we have to solve are

$$_{(p+1)}\eta^i_j e^j_p = 0 \qquad (i = 1, \ldots, \alpha_{p+1}), \qquad (5)$$

$$a^i_j e^j_p = \nu^i \qquad (i = 1, \ldots, R_p), \qquad (6)$$

where

$$\Gamma^i_p = a^i_j E^j_p.$$

We prove that the equations (5) and (6) have a solution by showing that the equations (5), (6) and the equations

$$_{(n-p+1)}\eta^{*i}_j e^j_p = 0 \qquad (i = 1, \ldots, \alpha_{p-1}), \qquad (7)$$

that is,

$$\sum_{j=1}^{\alpha_p} {}_{(p)}\eta^j_i e^j_p = 0 \qquad (i = 1, \ldots, \alpha_{p-1}),$$

have a unique solution.

The $\alpha_{p+1}+R_p+\alpha_{p-1}$ equations (5), (6) and (7) are not all independent, in general. There are exactly ρ_{p+1} independent equations (5), and ρ_p independent equations (7), where ρ_p is the rank of the matrix ${}_{(p)}\eta$. We have therefore at most

$$\rho_{p+1}+R_p+\rho_p = \alpha_p$$

linear equations to solve. The equations have a unique solution if the rank of the matrix of coefficients of the unknowns e_p^i is equal to α_p.

We have seen that the rank of the matrix of coefficients cannot exceed α_p. If the rank is less than α_p, the equations (5), (7) and

$$a_j^i e_p^j = 0 \qquad (i=1, \ldots, R_p) \qquad (6)'$$

have a solution in which the e_p^i are integers, not all zero. Suppose such a solution $(e_p^1, \ldots, e_p^{\alpha_p})$ exists.

Since the p-set $(e_p^1, \ldots, e_p^{\alpha_p})$ satisfies (5) and (6)$'$, it is closed and has all its periods zero. It is therefore a derived set. Hence there is a $(p-1)$-set $(e_{p-1}^1, \ldots, e_{p-1}^{\alpha_{p-1}})$ such that

$$e_p^i = {}_{(p)}\eta_j^i \, e_{p-1}^j.$$

Multiply the equations (7) by e_{p-1}^i, and apply the summation convention. We have

$$\sum_{j=1}^{\alpha_p} e_{p-1}^i \, {}_{(p)}\eta_i^j \, e_p^j = 0,$$

that is

$$\sum_{j=1}^{\alpha_p} (e_p^j)^2 = 0.$$

But the numbers e_p^j are integers, not all of which are zero. Hence we have a contradiction, and our assumption, that the matrix of coefficients of e_p^j in (5), (6) and (7) was of rank less than α_p, was false.

The rank of the matrix of coefficients is therefore exactly α_p, and equations (5), (6) and (7) have a solution, whatever values we give to ν^i. When we are only concerned with equations (5) and (6), we see that any solution of (5) and (6) is obtained by adding to the solution just found a solution of

(5) and (6)′. Thus the most general closed p-set which has assigned periods on R_p independent p-cycles of K is obtained by adding a derived set to a particular set having the required properties.

24·1. **Proof of de Rham's first theorem.** We are now in a position to prove de Rham's first theorem. Since the period with respect to a cycle Γ_p of the integral of a closed p-form on a manifold M is unaltered when Γ_p is replaced by a homologous cycle, we may without loss of generality assume that the base Γ_p^i $(i=1, \ldots, R_p)$ for the p-cycles of M is formed by cycles of a complex K which covers M. In order to find a regular closed form ϕ whose integral has the period ν^i on Γ_p^i, we first construct a closed set $(e_p^1, \ldots, e_p^{\alpha_p})$ which has the period ν^i on Γ_p^i $(i=1, \ldots, R_p)$. We then try to construct the regular closed form ϕ so that

$$\int_{E_p^i} \phi = e_p^i \qquad (i=1, \ldots, \alpha_p).$$

This is done by constructing regular p-forms

$$P_p^i \quad (i=1, \ldots, \alpha_p;\ p=0, \ldots, n)$$

with the properties:

(i) $$\int_{E_p^j} P_p^i = \delta_j^i;$$

(ii) $$(-1)^{p+1} P_{p,x}^i = {}_{(n-p)}\eta_j^{*i}\, P_{p+1}^j.$$

Suppose, indeed, that we have such a set of forms P_p^i. Then, if $(e_p^1, \ldots, e_p^{\alpha_p})$ is any p-set, we associate with it the p-form

$$\phi = \sum_{i=1}^{\alpha_p} e_p^i\, P_p^i.$$

Then
$$\begin{aligned}\phi_x &= \sum_{i=1}^{\alpha_p} e_p^i\, P_{p,x}^i \\ &= \sum_{i=1}^{\alpha_p} (-1)^{p+1}\, e_p^i\, {}_{(n-p)}\eta_j^{*i}\, P_{p+1}^j \\ &= \sum_{j=1}^{\alpha_{p+1}} {}_{(p+1)}\eta_i^j\, e_p^i\, P_{p+1}^j\end{aligned}$$

is the form associated with the derived $(p+1)$-set. If the given p-set is closed, ϕ is closed. Further, if $\Gamma_p = a_i E^i_p$ is any p-cycle,

$$\int_{\Gamma_p} \phi = \sum_{j=1}^{\alpha_p} a_j \int_{E^j_p} \sum_{i=1}^{\alpha_p} e^i_p P^i_p$$

$$= \sum_{i=1}^{\alpha_p} \sum_{j=1}^{\alpha_p} a_j e^i_p \delta^i_j$$

$$= a_j e^j_p,$$

and hence the period of the integral of ϕ on the cycle Γ_p is equal to the period of the closed p-set on Γ_p.

The proof of de Rham's first theorem is therefore reduced to the proof of the existence of the forms P^i_p with the properties (i) and (ii). It may be pointed out that the relation between the additive group of p-chains of K, and the additive group of p-forms $a_i P^i_p$ (where the coefficients a_i are real numbers), is that of a *group-pair*. Any p-chain C_p, and any p-form $a_i P^i_p$, determine a real number, viz.

$$\int_{C_p} a_i P^i_p,$$

with the properties:

(a) $$\int_{C_p} (a_i P^i_p + b_i P^i_p) = \int_{C_p} a_i P^i_p + \int_{C_p} b_i P^i_p,$$

and (b) $$\int_{C_p + C'_p} a_i P^i_p = \int_{C_p} a_i P^i_p + \int_{C'_p} a_i P^i_p.$$

The problem with which we are concerned therefore deals with a special case of the wider theory of group-pairs, which is of importance in topology. For an account of this theory, see Whitney [3].

24·2. As an aid to establishing the existence of the forms P^i_p, we first prove a lemma. In the number space $(x_1, \ldots, x_n)$ the set of points given by

$$|x_i| < 1 \qquad (i = 1, \ldots, n)$$

is called a *box*. We denote it by B_n. The lemma states:

Let P be a p-form ($p \geqslant 1$) which is regular in a region of a number space $(x_1, \ldots, x_n)$ containing the box B_n and which is zero outside the box. If P has the properties:

(i) *P is closed;*

(ii) *if $p = n$,*
$$\int_{B_n} P = 0,$$
then there exists a $(p-1)$-form Q, which is regular in the same region and vanishes outside B_n, such that
$$Q_x = P.$$
Moreover, if P depends on a parameter t and has the properties:

(*a*) *the rth derivatives of the coefficients of P with respect to t are continuous functions;*

(*b*) *P is zero when $|t| > k$;*

then Q can be chosen to satisfy (*a*) *and* (*b*).

It is worth while pointing out the reason for the condition (ii) which is imposed when $p = n$. If C_p is any p-chain whose boundary lies on the boundary $F(B_n)$ of B_n, C_p is homologous to a chain on $F(B_n)$ when p is less than n. Since P is closed, it follows that
$$\int_{C_n} P = 0.$$
But if $p = n$, $C_n = B_n$ is a chain whose boundary lies on $F(B_n)$, and C_n is not homologous to a chain on $F(B_n)$. We cannot therefore deduce from the fact that P is closed that
$$\int_{C_n} P = 0,$$
and condition (ii) is necessary to ensure this result.

We prove our lemma by induction on n, the dimension of the number space. If $n = 1$, we have only one case to consider, namely $p = 1 = n$. B_n is given by
$$|x| < 1.$$

If P is a 1-form satisfying (i) and (ii), and

$$Q = \int_{-1}^{x} P,$$

Q is regular and vanishes outside B_1, on account of (ii). Further,

$$Q_x = P.$$

The form Q therefore fulfils the conditions of our lemma. Moreover, if P depends on a parameter t and satisfies the supplementary conditions (a) and (b), Q also satisfies these conditions.

We therefore assume that our lemma has been established for space of less than n dimensions, and show that this implies its truth for a space of n dimensions. We consider first the case $p = n$. Let

$$P = A\, dx^1 \ldots dx^n$$

and write

$$P^1 = A\, dx^1 \ldots dx^{n-1}.$$

Let $[x_n]$ denote the box of $(n-1)$ dimensions which is the section of B_n by x_n = constant, and write

$$\int_{[x_n]} P^1 = f(x_n), \quad \int_{-1}^{x_n} f(x_n)\, dx^n = F(x_n).$$

We can construct an $(n-1)$-form of Q^1, which is independent of x_n, with the properties:

(1) Q^1 is regular in a region of our space containing B_n;

(2) Q^1 is zero outside $[x_n]$ in the $(n-1)$-space x_n = constant;

(3) $$\int_{[x_n]} Q^1 = 1.$$

Then, if we regard the $(n-1)$-form

$$P^1 - f(x_n) \times Q^1$$

as a form in $[x_n]$ containing the parameter x_n, it follows from the hypothesis of induction that there exists an $(n-2)$-form Q^2 with the properties:

(1) Q^2 is regular in our region;

(2) Q^2 is zero outside $[x_n]$ in the $(n-1)$-space $x_n =$ constant;

(3) $$Q^2_x = P^1 - f(x_n) \times Q^1 \quad \text{in} \quad [x_n];$$

(4) when regarded as a form in the n-space, Q^2 is regular and vanishes outside B_n.

This last property follows from the supplementary conditions.

We now regard Q^2 as a form in the n-space. We have

$$\begin{aligned}[Q^2 \times dx^n]_x &= P - f(x_x) \times Q^1 \times dx^n \\ &= P + (-1)^n [F(x_n) \times Q^1]_x.\end{aligned}$$

If, now, we define the form Q by the equation

$$Q = Q^2 \times dx^n + (-1)^{n-1} F(x_n) \times Q^1,$$

Q satisfies all the requirements of our lemma. If P contains a parameter t and satisfies the supplementary conditions (a) and (b), Q also satisfies (a) and (b), provided Q^1 and Q^2 satisfy the condition relating to the parameter t.

We now come to the case $p < n$. Let P^1 be the form obtained by putting $dx^n = 0$ in P. We write

$$P = P^1 + P^2 \times dx^n.$$

The form P^1 is closed in $[x_n]$, and, if $p = n-1$,

$$\begin{aligned}\int_{[x_n]} P^1 &= \int_{[x_n]} P \\ &= \int_{\Gamma} P = 0,\end{aligned}$$

where Γ is the boundary of one of the cells into which $[x_n]$ divides A. From our induction hypothesis, using the supplementary conditions, we know that there exists a form Q^1

which is regular in the region containing B_n and is zero outside B_n, such that

$$Q^1_x = P^1$$

when $dx^n = 0$. Thus we can write

$$Q^1_x = P^1 + P^3 \times dx^n,$$

and hence $$P - Q^1_x = (P^2 - P^3) \times dx^n.$$

Since $P - Q^1_x$ is closed, $P^2 - P^3$ must be closed, when regarded as a form in $[x_n]$. It also vanishes outside $[x_n]$ in $x_n = \text{constant}$. Hence there exists a form Q^2, which is regular in our region and vanishes outside B_n, such that

$$Q^2_x = P^2 - P^3$$

in $[x_n]$. Hence, in the n-space,

$$[Q^2 \times dx^n]_x = P - Q^1_x$$

and so $$Q = Q^1 + Q^2 \times dx^n$$

is a $(p-1)$-form satisfying the requirements of our lemma. If P depends on a parameter t and satisfies the supplementary conditions (*a*) and (*b*), it is easily seen that Q also satisfies conditions (*a*) and (*b*), provided Q^1 and Q^2 are chosen to satisfy these conditions.

24·3. We now return to our manifold M of n dimensions, with a covering complex K. In order to construct the forms P^i_p, we construct a series of domains on M. Each of these domains is in (1-1) continuous correspondence of class v with a box B_n, and if v is at least *three* any regular form in the box will correspond to a regular form in the domain.

If E^i_p is any simplex of K, the region of M which is covered by E^i_p and the simplexes of K which have E^i_p on their boundaries is called the *star* of E^i_p in K. We construct a domain associated with each cell of the dual complex K^*. The domain associated with E^{*i}_{n-p} is denoted by $D(E^{*i}_{n-p})$, and the domains are required to satisfy the conditions:

(1) each domain is in (1-1) correspondence of class v ($v \geqslant 3$) with a box B_n;

(2) $D(E^{*i}_{n-p})$ lies in the star of E^i_p and contains E^{*i}_{n-p} in its interior;

(3) if $E^{*j_k}_{n-p-1}$ $(k=1, \ldots, r)$ are the $(n-p-1)$-cells of K^* which lie on the boundary of E^{*i}_{n-p}, the domains $D(E^{*j_k}_{n-p-1})$ $(k=1, \ldots, r)$ are contained in $D(E^{*i}_{n-p})$.

It is easily shown that it is possible to construct such a set of domains.

We now prove that there exist regular forms P^i_p $(i=1, \ldots, \alpha_p;\ p=0, \ldots, n)$ satisfying the three conditions:

(i) $$\int_{E^j_p} P^i_p = \delta^i_j;$$

(ii) $$(-1)^{p+1} P^i_{p,x} = {}_{(n-p)}\eta^{*i}_j P^j_{p+1};$$

(iii) P^i_p is zero outside $D(E^{*i}_{n-p})$.

The forms P^i_n are easily obtained. We can construct a function $f_i(x)$ on M, of class *two* at least, which vanishes outside $D(E^{*i}_0)$ and satisfies the condition

$$\int_{D(E^{*i}_0)} f_i(x)\, dx^1 \ldots dx^n = 1.$$

We define P^i_n to be $f_i(x)\, dx^1 \ldots dx^n$. Since $D(E^{*i}_0)$ lies in E^i_n, P^i_n satisfies conditions (i) and (iii). Condition (ii) does not arise.

To construct P^i_{n-1}, let

$$E^{*i}_1 \to E^{*j}_0 - E^{*k}_0.$$

Then $P^j_n - P^k_n$ vanishes outside $D(E^{*i}_1)$, and

$$\int_{D(E^{*i}_1)} (P^j_n - P^k_n) = 0.$$

Hence, applying the lemma of the last paragraph, we can construct an $(n-1)$-form P^i_{n-1}, regular on M and vanishing outside $D(E^{*i}_1)$, which satisfies the equation

$$(-1)^n P^i_{n-1,x} = P^j_n - P^k_n.$$

This form P^i_{n-1} satisfies the conditions (ii) and (iii), by construction. It also satisfies condition (i) when $i \neq j$. To show that it satisfies this condition when $i = j$, we consider the boundary relation for E^j_{n-1}. This is

$$E^j_n \to (-1)^n E^i_{n-1} + C_{n-1},$$

where C_{n-1} is a chain of K in which E^i_{n-1} has the coefficient zero. Then

$$\begin{aligned}\int_{E^i_{n-1}} P^i_{n-1} &= (-1)^n \int_{F(E^j_n)} P^i_{n-1} \\ &= \int_{E^j_n} (P^j_n - P^k_n) \\ &= 1.\end{aligned}$$

The forms P^i_p are obtained inductively. Suppose we have constructed P^i_q, for $q > p$. Then

$$\begin{aligned}[(-1)^{p+1}\,{}_{(n-p)}\eta^{*i}_j P^j_{p+1}]_x &= -{}_{(n-p)}\eta^{*i}_j\,{}_{(n-p-1)}\eta^{*j}_k P^k_{p+2} \\ &= 0,\end{aligned}$$

and $(-1)^{p+1}\,{}_{(n-p)}\eta^{*i}_j P^j_{p+1}$ is regular and vanishes outside $D(E^{*i}_{n-p})$. Hence, by our lemma, there exists a p-form P^i_p regular on M and vanishing outside $D(E^{*i}_{n-p})$, such that

$$P^i_{p,x} = (-1)^{p+1}\,{}_{(n-p)}\eta^{*i}_j P^j_{p+1}.$$

The form P^i_p satisfies conditions (ii) and (iii), and condition (i) when $i \neq j$, by construction. It remains to show that condition (i) is satisfied when $i = j$. In the following equations the summation convention is *not* used. We have

$$\int_{E^i_p} P^i_{p,x} = {}_{(p+1)}\eta^k_i \int_{F(E^k_{p+1})} P^i_p,$$

where E^k_{p+1} is a simplex of K having E^i_p on its boundary. Hence

$$\begin{aligned}\int_{E^i_p} P^i_p &= {}_{(p+1)}\eta^k_i \int_{E^k_{p+1}} P^i_{p,x} \\ &= {}_{(p+1)}\eta^k_i \left[(-1)^{p+1} \sum_{j=1}^{\alpha_{p+1}} {}_{(n-p)}\eta^{*i}_j \int_{E^k_{p+1}} P^j_{p+1}\right] \\ &= [{}_{(p+1)}\eta^k_i]^2 = 1,\end{aligned}$$

since $\qquad {}_{(p+1)}\eta^k_i = (-1)^{p+1}\,{}_{(n-p)}\eta^i_k = \pm 1.$

In this way all the forms P_p^i are constructed. The truth of the first theorem of de Rham follows at once. Thus, we have finally proved that there exists a closed p-form whose integral has R_p arbitrarily assigned real numbers as its periods on R_p independent p-cycles of M.

25. De Rham's second theorem. It is appropriate at this stage to mention the second theorem of de Rham. This theorem states that *if ϕ is a closed form on a manifold M whose integral has all its periods equal to zero, then ϕ is a null form.* De Rham has proved this theorem by a method which is similar to that used above to establish the first theorem. Let ϕ be a closed p-form whose integral has zero periods. We define a p-set $(e_p^1, \ldots, e_p^{\alpha_p})$ by the equations

$$e_p^i = \int_{E_p^i} \phi.$$

This p-set is closed, and has all its periods equal to zero. Hence, by § 23·2, it is a derived set, and we can write

$$e_p^i = {}_{(p)}\eta_j^i e_{p-1}^j.$$

Now
$$\begin{aligned}\sum_{i=1}^{\alpha_p} e_p^i P_p^i &= \sum_{i=1}^{\alpha_p} {}_{(p)}\eta_j^i e_{p-1}^j P_p^i \\ &= \sum_{j=1}^{\alpha_{p-1}} (-1)^p {}_{(n-p+1)}\eta_i^{*j} e_{p-1}^j P_p^i \\ &= \left[\sum_{j=1}^{\alpha_{p-1}} e_{p-1}^j P_{p-1}^j\right]_x.\end{aligned}$$

Hence
$$\phi \sim \psi = \phi - \sum_{i=1}^{\alpha_p} e_p^i P_p^i,$$

where ψ is a closed form which has all its periods zero, and which has the further property that

$$\int_{E_p^i} \psi = 0 \qquad (i = 1, \ldots, \alpha_p). \tag{8}$$

To establish de Rham's second theorem, it is sufficient to prove that ψ is a null form. While there is no intrinsic difficulty in proving this, the details of the proof are somewhat involved. We have first to extend the lemma of § 24·2, and we have then to construct a new and elaborate system of domains on M. De Rham first shows that ψ is homologous to a p-form ψ^1, which has the property (8), and which vanishes in a set of domains containing the 0-simplexes of K. Then he proves that ψ^1 is homologous to a p-form ψ^2 which has the property (8) and which vanishes in a set of domains containing the 1-simplexes of K. Proceeding by induction, he proves that ψ is homologous to a p-form ψ^{n+1}, where ψ^{n+1} vanishes in a set of domains containing the n-simplexes of K. But this last set of domains covers M, and ψ^{n+1} is therefore zero on M. The theorem follows.

We do not go into the details of this proof, which will be found in de Rham's paper [2]. Instead, we shall prove the theorem in the next chapter, as a first application of the properties of harmonic integrals.

26·1. **Products of integrals and intersections of cycles.** The two theorems of de Rham enable us to express some of the topological invariants of a manifold M in terms of the integrals on M. If Γ_p^i $(i = 1, \ldots, R_p)$ is a base for the p-cycles of M, the first theorem shows that there exist closed p-forms ϕ_p^i $(i = 1, \ldots, R_p)$ such that

$$\int_{\Gamma_p^j} \phi_p^i = \delta_j^i.$$

No linear combination with constant coefficients of the p-forms ϕ_p^i can be a null form. For, if

$$\phi = a_i \phi_p^i \sim 0,$$

the integral of ϕ must have all its periods zero. But

$$\int_{\Gamma_p^i} \phi = a_i,$$

and hence each coefficient a_i is zero. Again, if ψ is any closed

p-form having the period ν^i on Γ_p^i $(i=1, \ldots, R_p)$, then $\psi - \nu_i \phi_p^i$ has all its periods zero. Hence, by de Rham's second theorem,

$$\psi \sim \sum_{i=1}^{R_p} \nu^i \phi_p^i.$$

The pth Betti number of M can therefore be defined as the maximum number of closed p-forms on M which are *linearly independent*, that is, which are such that no linear combination of them is a null form.

This result suggests that we might try to express all the theorems on the topology of a manifold in terms of properties of integrals on the manifold. This procedure is not, however, completely satisfactory. In the first place, we have used the properties of complexes on a manifold in order to establish the existence of closed forms with non-zero periods. While it might be possible to overcome this difficulty, there is a second and more serious objection. Certain topological invariants, such as the torsion groups of a manifold, have no place in the theory of integrals, and cannot be taken account of in an investigation of the properties of a manifold by means of integrals. Nevertheless, when we only need to consider properties of a manifold depending on homology with division, the topological characters with which we are chiefly concerned can be expressed in terms of integrals. We now give some of the more important formulae.

26·2. Let Γ_p^i $(i=1, \ldots, R_p)$ be a base for the p-cycles of M, for $p = 0, \ldots, n$, and let ϕ_p^i denote a closed p-form on M such that

$$\int_{\Gamma_p^j} \phi_p^i = \delta_j^i.$$

A set of closed p-forms ψ_p^i $(i=1, \ldots, R_p)$ is said to form a base for the p-forms on M if any closed p-form ψ on M satisfies a homology

$$\psi \sim a_i \psi_p^i,$$

where the coefficients a_i are real numbers. The result stated in § 26·1 shows that the forms ϕ_p^i $(i=1,\ldots,R_p)$ form a base. We now show that the necessary and sufficient condition that the closed forms ψ_p^i $(i=1,\ldots,R_p)$ should form a base is that the period matrix $\boldsymbol{\omega} = (\omega^{ij})$ should be non-singular, where

$$\omega^{ij} = \int_{\Gamma_p^j} \psi_p^i.$$

(i) The condition is necessary. For, if the forms ψ_p^i form a base, there exist real numbers a_j^i such that

$$\phi_p^i \sim a_j^i \psi_p^j.$$

Hence
$$\delta_k^i = a_j^i \omega^{jk}.$$

Since the matrix formed by the first members of these equations is the unit matrix $\mathbf{I}_{R_p}$, (a_j^i) and (ω^{ij}) are non-singular matrices.

(ii) The condition is sufficient. If (ω^{ij}) is a non-singular matrix, it possesses an inverse, which we denote by (a^{ij}). Then, if

$$\chi_p^i = \sum_{j=1}^{R_p} a^{ij} \psi_p^j,$$

the integral of $\chi_p^i - \phi_p^i$ has all its periods zero, and hence

$$\chi_p^i \sim \phi_p^i.$$

If ψ is any closed p-form on M,

$$\begin{aligned} \psi &\sim b_i \phi_p^i \\ &\sim \sum_{j=1}^{R_p} b_i a^{ij} \psi_p^j, \end{aligned}$$

where the coefficients b_i are real numbers.

The condition that a set of closed p-forms should form a base has been expressed in terms of the period matrix of the forms. This therefore requires the choice of a base for the p-cycles. We can express the condition without any reference to the periods. Let ψ_{n-p}^i $(i=1,\ldots,R_{n-p})$ be a set of closed $(n-p)$-forms forming a base for the $(n-p)$-forms on M, and let the period matrix of their integrals be $\boldsymbol{\nu} = (\nu^{ij})$. Then, by the bilinear relations,

$$\left\| \int_M \psi_p^i \times \psi_{n-p}^j \right\| = \boldsymbol{\omega}\boldsymbol{\epsilon}^p\boldsymbol{\nu}',$$

where $\boldsymbol{\epsilon}^p$ is the transpose of the inverse of the intersection matrix $\| (\Gamma_p^i \, . \, \Gamma_{n-p}^j) \|$. The matrices $\boldsymbol{\epsilon}^p$ and $\boldsymbol{\nu}$ are non-singular, and hence $\left\| \int_M \psi_p^i \times \psi_{n-p}^j \right\|$ is non-singular if and only if $\boldsymbol{\omega}$ is non-singular. Hence:

A set of closed p-forms ψ_p^i $(i = 1, \ldots, R_p)$ is a base for the p-forms on M if and only if there exists a set of closed $(n-p)$-forms ψ_{n-p}^i $(i = 1, \ldots, R_{n-p})$ such that the matrix

$$\left\| \int_M \psi_p^i \times \psi_{n-p}^j \right\|$$

is non-singular.

26·3. Let Γ_p^i $(i = 1, \ldots, R_p)$ be a base for the p-forms on M, defined for each value of p. If ϕ_p^i $(i = 1, \ldots, R_p)$ is a base for the p-cycles of M, we now seek relations connecting the numbers

$$\int_M \psi_p^i \times \psi_q^j \qquad (p + q = n),$$

$$\int_M \psi_a^i \times \psi_b^j \times \psi_c^k \qquad (a + b + c = n),$$

etc., with the intersection numbers

$$(\Gamma_{n-p}^i \, . \, \Gamma_{n-q}^j), \quad (\Gamma_{n-a}^i \, . \, \Gamma_{n-b}^j \, . \, \Gamma_{n-c}^k), \quad \ldots.$$

These intersection numbers depend on the choice of the bases for the cycles of M, so we must expect the relations to depend on the periods of the integrals of the forms ϕ_p^i with respect to the cycles Γ_p^i. The formulae are simplest when we choose the forms ϕ_p^i so that

$$\int_{\Gamma_p^j} \phi_p^i = \delta_j^i \qquad (p = 0, \ldots, n),$$

and we shall consider only this case, leaving the deduction of the formulae when the bases ϕ_p^i $(i=1, \ldots, R_p)$ are chosen generally as an exercise to the reader.

We denote the inverse of the matrix $\| (\Gamma_{n-p}^i \cdot \Gamma_p^j) \|$ by $\mathbf{A}^p$.

(i) Let $p+q=n$. Then, if

$$\int_M \phi_p^i \times \phi_q^j = k_{ij}^{pq},$$

we know, by the bilinear relations, that the matrix $\mathbf{k}^{pq}$ is the inverse of the transpose of the matrix $\| (\Gamma_p^i . \Gamma_q^j) \|$. The transpose of this intersection matrix is, however, equal to $(-1)^{pq} (\mathbf{A}^p)^{-1}$, and hence

$$\mathbf{k}^{pq} = (-1)^{pq} \mathbf{A}^p.$$

(ii) Let $p+q+r=n$. The periods of the integral of the $(p+q)$-form $\phi_p^i \times \phi_q^j$ are found by the formula of § 22·5. We have

$$\int_{\Gamma_{n-r}^\gamma} \phi_p^i \times \phi_q^j = \lambda_{ij}^{n-q}(\Gamma_{n-r}^\gamma),$$

and hence $$\phi_p^i \times \phi_q^j \sim \sum_{\gamma=1}^{R_r} \lambda_{ij}^{n-q}(\Gamma_{n-r}^\gamma)\, \phi_{n-r}^\gamma.$$

Therefore, we have

$$k_{ijk}^{pqr} = \int_M \phi_p^i \times \phi_q^j \times \phi_r^k = \int_M \sum_{\gamma=1}^{r} \lambda_{ij}^{n-q}(\Gamma_{n-r}^\gamma) \int_M \phi_{n-r}^\gamma \times \phi_r^k$$
$$= (-1)^{r(n-r)} \lambda_{ij}^{n-q}(\Gamma_{n-r}^\gamma)\, A_{\gamma k}^{n-r}.$$

If we now substitute the value of $\lambda_{ij}^{n-q}(\Gamma_{n-r}^\gamma)$ given by equation (12) of § 18·3, we obtain the equation

$$k_{ijk}^{pqr} = A_{\alpha i}^{n-p} A_{\beta j}^{n-q} A_{\gamma k}^{n-r} (\Gamma_{n-p}^\alpha \cdot \Gamma_{n-q}^\beta \cdot \Gamma_{n-r}^\gamma).$$

A similar argument enables us to express quadruple intersection numbers in terms of integrals over M of quadruple products of the forms ϕ_p^i, etc.

In the case in which n is a multiple of 4, say $n=2m$ where m is even, we have

$$\mathbf{k}^{mm} = \mathbf{A}^{mm},$$

where $\mathbf{A}^{mm}$ is the inverse of the intersection matrix $\| (\Gamma_m^i . \Gamma_m^j) \|$. The signature of this intersection matrix is therefore equal to the signature of $\mathbf{k}^{mm}$. Let ψ_m^i $(i=1, \ldots, R_m)$ be any base for the m-cycles of M, and let

$$\int_{\Gamma_m^j} \psi_m^i = \omega^{ij}.$$

Then

$$\left\| \int_M \psi_m^i \times \psi_m^j \right\| = \boldsymbol{\omega} \mathbf{k}^{mm} \boldsymbol{\omega}',$$

and it follows that the signature of $\left\| \int_M \psi_m^i \times \psi_m^j \right\|$ is equal to the signature of $\| (\Gamma_m^i . \Gamma_m^j) \|$. Hence the signature of

$$\left\| \int_M \psi_m^i \times \psi_m^j \right\|$$

is a topological invariant of M.

REFERENCES

1. E. Goursat. *Cours d'Analyse*, vol. 1, 4th ed. (Paris), 1924.
2. G. de Rham. *Journal de Mathématique* (9), 10 (1931), 115.
3. H. Whitney. *Duke Mathematical Journal*, 3 (1937), 35.

Chapter III

HARMONIC INTEGRALS

The investigation of the properties of integrals on a manifold which we have made in Chapter II does not make any use of the Riemannian metric which was introduced in Chapter I. In this chapter we consider integrals in relation to a given metric, and use the fundamental metrical tensor in order to define a restricted class of regular closed forms. This class of closed forms has the property that there is in it exactly one form of multiplicity p homologous to any regular closed form of multiplicity p, or, in other words, there is just one p-form of the class whose integral has given periods on independent p-cycles of the manifold. The definition of this class of forms has certain analogies with the definition of potential functions in mathematical physics, or of harmonic functions in the theory of functions of a complex variable, and for this reason we call the forms of the class *harmonic forms*, and their integrals *harmonic integrals*; we also call the skew-symmetric covariant tensor, defined by the coefficients of a harmonic integral, a *harmonic tensor*.

27·1. **Definition of harmonic forms.** Let us consider a Euclidean space of n-dimensions, in which $(x_1, \ldots, x_n)$ are rectangular cartesian coordinates. From this space we can define a Riemannian space in the following way. The Riemannian space has the same points and neighbourhoods as the Euclidean space, but we now admit as allowable local coordinate systems any set of coordinates $(y_1, \ldots, y_n)$ obtained from the cartesian coordinates $(x_1, \ldots, x_n)$ by a transformation of class u $(u \geqslant 2)$ which satisfies conditions (i) and (ii) of § 2·1, in some region. We define the metric by the condition that in the coordinate system $(x_1, \ldots, x_n)$ it is given by

$$g_{ij}dx^i dx^j = (dx^1)^2 + \ldots + (dx^n)^2.$$

Let ϕ be a function in the Euclidean space which is a potential function in some region of the space. The condition for this is, in the usual vector notation,

$$\mathbf{div}\ \mathbf{grad}\ \phi = 0.$$

Now **grad** ϕ is, in our notation, the covariant vector

$$\phi_i = \frac{\partial \phi}{\partial x_i}.$$

Thus when the space is considered as a Riemannian space, **grad** ϕ is the covariant derivative of ϕ. The equation

$$\mathbf{div}\ \phi_i = 0$$

is, when written in full,

$$\Sigma \frac{\partial \phi_i}{\partial x_i} = 0,$$

and in this form it is not a tensor equation in the Riemannian space. We now try to find a tensor equation in the Riemannian space which will reduce to **div** $\phi_i = 0$ when the coordinates are the rectangular cartesians $(x_1, \ldots, x_n)$.

In the coordinate system $(x_1, \ldots, x_n)$ we have the relations

$$g_{ij} = g^{ij} = \delta^i_j, \qquad \Gamma^i_{jk} = 0.$$

Moreover, we have $\sqrt{g} = 1$. The equation **div** $\phi_i = 0$ can therefore be written as

$$\frac{1}{\sqrt{g}} \sum_{i=1}^{n} \frac{\partial}{\partial x_i} (\sqrt{g}\, g^{ij} \phi_j) = 0,$$

or, equivalently, $$g^{ij}\phi_{j,i} = 0.$$

This last equation is a tensor equation in the Riemannian space, and is the necessary and sufficient condition that the scalar field ϕ should correspond to a potential function in the

Euclidean space. If we now define the divergence of an absolute covariant vector ξ_i in the Riemannian space by the equation

$$\mathbf{div}\,\xi_i = g^{ij}\xi_{i,j},$$

Laplace's equation $\mathbf{div}\,\mathbf{grad}\,\phi = 0$ becomes a tensor equation in the Riemannian space. A function ϕ which satisfies it may be called a harmonic function, and the necessary and sufficient condition that a function ϕ should be a potential (or harmonic) function in the Euclidean space is that it should be harmonic in the Riemannian space.

The equation $$\frac{1}{\sqrt{g}}\sum_{i=1}^{n}\frac{\partial}{\partial x_i}(\sqrt{g}\,g^{ij}\phi_j) = 0,$$

which we found above for the condition that ϕ should be harmonic, is the condition that the $(n-1)$-form

$$\frac{1}{(n-1)!}\sqrt{g}\,g^{ij}\phi_j\epsilon_{i\,i_1\ldots i_{n-1}}dx^{i_1}\ldots dx^{i_{n-1}}$$

should be closed. This geometrical form of the condition suggests the generalisation of the notion of a harmonic function which we are seeking.

27·2. We consider a general n-dimensional Riemannian space, with the metrical tensor g_{ij}. Let

$$P = \frac{1}{p!}P_{i_1\ldots i_p}dx^{i_1}\ldots dx^{i_p}$$

be a p-form in the space. Since $P_{i_1\ldots i_p}$ is an absolute tensor,

$$P^*_{i_1\ldots i_{n-p}} = \frac{1}{p!}\sqrt{g}\,g^{j_1k_1}\ldots g^{j_pk_p}P_{k_1\ldots k_p}\epsilon_{j_1\ldots j_p\,i_1\ldots i_{n-p}}$$

is a skew-symmetric tensor of weight zero, and the form

$$P^* = \frac{1}{(n-p)!}P^*_{i_1\ldots i_{n-p}}dx^{i_1}\ldots dx^{i_{n-p}}$$

which it defines is determined uniquely by P and the metrical tensor. We call P^* the *dual* of the form P. Now

$$P^{**}_{i_1\dots i_p} = \frac{1}{(n-p)!}\sqrt{g}\, g^{j_1k_1}\dots g^{j_{n-p}k_{n-p}}\, P^*_{k_1\dots k_{n-p}}\, \epsilon_{j_1\dots j_{n-p}\, i_1\dots i_p}$$

$$= \frac{1}{p!(n-p)!} g\, g^{j_1k_1}\dots g^{j_{n-p}k_{n-p}}\, g^{a_1b_1}\dots g^{a_pb_p}$$

$$\times P_{b_1\dots b_p}\, \epsilon_{a_1\dots a_p\, k_1\dots k_{n-p}}\, \epsilon_{j_1\dots j_{n-p}\, i_1\dots i_p};$$

using geodesic coordinates, we can easily prove that

$$P^{**}_{i_1\dots i_p} = (-1)^{p(n-p)}\, P_{i_1\dots i_p}.$$

In these coordinates, we have, at the origin,

$$P^{**}_{i_1\dots i_p} = (-1)^{p(n-p)}\, P^*_{i_{p+1}\dots i_n}$$

$$= (-1)^{p(n-p)}\, P_{i_1\dots i_p}.$$

Hence we have $\qquad P^{**} = (-1)^{p(n-p)}\, P,$

which shows that the relation between P and P^* is symmetrical, save perhaps for sign. If P is the exterior derivative Q_x of a $(p-1)$-form Q, we shall usually write

$$P^* = Q^x.$$

If P is an 0-form, that is a function, Laplace's equation can be written as

$$[P^x]_x = P^x{}_x = 0.$$

An example of a 1-form which satisfies the equation

$$P^x{}_x = 0$$

suggests itself at once. In Euclidean space of three dimensions, consider the magnetic field due to a system of currents. At any point of the space at which there is no current, the vector potential $\mathbf{A} = A_i$ is a covariant vector satisfying the equation

$$\mathbf{curl\ curl\, A} = 0.$$

If we now pass from the Euclidean space to the Riemannian space defined by it, as in § 27·1, and consider the 1-form $A = A_i dx^i$, this vector equation can be written as

$$A^x{}_x = 0.$$

27·3. It will be remembered that the electrical potential due to a system of charges, or the vector potential due to a system of currents, is not uniquely determined; to the former we may add an arbitrary constant and to the latter we may add a vector whose **curl** is zero, that is, a tensor defining a 1-form which is closed. Let us consider briefly the usual method of defining the potential or vector potential. In defining electrical potential we start with a covariant vector **E**, representing the electrical intensity, which satisfies the equation

$$\mathbf{curl}\,\mathbf{E} = 0,$$

or the equivalent equation

$$E = E_i dx^i \rightarrow 0.$$

We then say that this is sufficient to define a function ϕ, unique save for an additive constant, which satisfies the equation

$$\phi_x = E.$$

Similarly, in defining the vector potential we begin with the contravariant vector $\mathbf{B} = B^i$ which represents the magnetic induction. This satisfies the equation

$$\mathbf{div}\,\mathbf{B} = 0,$$

and the equivalent equation in the notation of forms is

$$B = B_{ij}\, dx^i dx^j \rightarrow 0,$$

where $$B_{ij} = \sqrt{g}\, \epsilon_{ijk} B^k.$$

We now say that this is a sufficient condition for the existence of a 1-form

$$A = A_i dx^i,$$

determined save for an additive closed 1-form, which satisfies

$$A_x = B.$$

But the theorem invoked, that a p-form is null if it is closed, is only true *locally*, that is, in a sufficiently small neighbourhood of a point, and, as we saw in Chapter II, it is not true in the large on a manifold for which R_p is greater than zero. We now define harmonic tensors to be the analogues of the electrical intensity and magnetic induction *in the large*, and we are thus led to the following definition: *A p-form P is a harmonic form if* (1) *it is regular everywhere on M, and* (2) *it satisfies everywhere the conditions*

$$P \to 0, \quad P^* \to 0.$$

The integral of a harmonic form is a *harmonic integral*, and the tensor defined by the coefficients of a harmonic form is a *harmonic tensor*. According to this definition the potential difference between two points in an electrical field is measured by the integral of a harmonic form along any path between the two points; and the integral

$$\int \mathbf{A} \, . \, \mathbf{dx}$$

of the vector potential in the magnetic field round a bounding circuit is equal to the integral of a harmonic 2-form over any 2-chain having the circuit as its boundary.

As immediate consequences of the definition we observe that the $(n-p)$-form P^* which is the dual of a harmonic p-form P is harmonic, and that a harmonic 0-form is a constant while a harmonic n-form is a constant multiple of

$$\sqrt{g}\, dx^1 \ldots dx^n.$$

27·4. It is not of course clear that on a Riemannian manifold M, p-forms P $(0 < p < n)$ having these properties exist, and one of our main objects in this chapter will be to establish the existence of a harmonic p-form P such that the integral

$$\int P$$

has arbitrarily assigned periods on R_p independent p-cycles of M. The case $p = 1$, $n = 2$ is, however, somewhat special, and the existence theorem in this case is really a classical theorem. The Riemannian manifold is in this case a two-sided closed surface (cf. § 1), and by elementary differential geometry we know that we can choose local systems of parameters (x, y) so that the fundamental quadratic differential form is

$$\lambda(x, y)\,[dx^2 + dy^2].$$

Now if

$$U = P\,dx - Q\,dy$$

is harmonic, then

$$\frac{\partial P}{\partial y} + \frac{\partial Q}{\partial x} = 0,$$

and

$$\frac{\partial P}{\partial x} = \frac{\partial Q}{\partial y},$$

so that if $U^* = V = P\,dy + Q\,dx$, the integral

$$\int (U + iV) = \int (P + iQ)\,(dx + i\,dy)$$

is locally an analytic function of the complex variable $z = x + iy$. It follows that, if we apply to M any of the classical proofs, e.g. that given by Weyl [7], for the existence of everywhere finite algebraic integrals on a Riemann surface, we establish the existence of a unique harmonic integral

$$\int (P\,dx - Q\,dy)$$

having assigned periods on the R_1 1-cycles of M.

28·1. **Approximation by closed p-sets.** It is not without interest to consider the relation between the harmonic integrals which we have defined and certain p-sets which we discussed in § 23. To do this, we first define a q-dimensional direction at a point P of a manifold. Let $(x_1, \ldots, x_n)$ be coordinates valid in a neighbourhood of P, and consider a locus defined by the equations

$$x_i = f_i(t_1, \ldots, t_q) \qquad (i = 1, \ldots, n),$$

which passes through P. We say that P is an *ordinary point* of the locus if the matrix $\left(\frac{\partial f_i}{\partial t_j}\right)$ is of rank q at P. If L' is another locus, given by

$$x_i = g_i(u_1, \dots, u_q) \qquad (i = 1, \dots, n),$$

which passes through P and has an ordinary point at P, we say that L and L' touch at P if

$$\left(\frac{\partial f_i}{\partial t_j}\right)_P = \left(\frac{\partial g_i}{\partial u_k}\right)_P a_j^k,$$

where (a_j^k) is a non-singular matrix of q rows and columns. By using elementary properties of functions we can verify that the following properties hold:

(1) the condition that L and L' should touch is independent of the parameters in L and L' which are used;

(2) the condition that L and L' should touch does not depend on the coordinate system $(x_1, \dots, x_n)$ chosen at P;

(3) if L touches L' at P and L' touches L'' at P, then L touches L'' at P.

The set of all loci which have an ordinary point at P and touch at P is said to define a *direction* at P.

Let us suppose that the point P is given by $t_i = 0$ $(i = 1, \dots, q)$ in the equation of the locus L. We consider the portion S of L given by

$$|t_i| < \alpha_i \qquad (i = 1, \dots, q)$$

whose $\alpha_1, \dots, \alpha_q$ are positive constants. The *volume* $V(S)$ of S is defined by the integral

$$\int_{-\alpha_1}^{\alpha_1} \cdots \int_{-\alpha_q}^{\alpha_q} \left\{\frac{1}{q!} g_{i_1 j_1} \cdots g_{i_q j_q} \frac{\partial(f_{i_1}, \dots, f_{i_q})}{\partial(t_1, \dots, t_q)} \frac{\partial(f_{j_1}, \dots, f_{j_q})}{\partial(t_1, \dots, t_q)}\right\}^{\frac{1}{2}} dt^1 \dots dt^q.$$

Let
$$Q = \frac{1}{q!} Q_{i_1 \dots i_q} dx^{i_1} \dots dx^{i_q}$$

be any q-form, defined in the neighbourhood of P. We consider the limit of

$$\frac{1}{V(S)}\int_S Q$$

as $\alpha_1, \ldots, \alpha_q$ tend to zero independently. By elementary analysis it can be shown that this limit depends only on Q and the direction of L at P. We define it as the *derivative* of the form Q in the given direction at P.

This definition leads to an interesting interpretation of the condition that a p-form P should be a harmonic form. Consider any point O of our manifold, and a p-dimensional direction through it. We can choose local coordinates valid in the neighbourhood of O, which are geodesic at O, and are such that the locus

$$x_{i_{p+1}} = 0, \ldots, x_{i_n} = 0$$

is tangent to the assigned direction at O. In these coordinates, the derivative of P in the given direction is $\pm P_{i_1 \ldots i_p}$ evaluated at O. Since the coordinates are geodesic, the dual form P^* has the property that

$$P^*_{i_{p+1} \ldots i_n} = P_{i_1 \ldots i_p},$$

provided $(i_1, \ldots, i_n)$ is an even derangement of the natural order. The direction absolutely perpendicular to the given direction is defined by the locus

$$x_{i_1} = 0, \ \ldots, \ x_{i_p} = 0,$$

and hence the derivative of P^* in this direction is $\pm P_{i_1 \ldots i_p}$ evaluated at O.

Hence the dual form P^* of P can be defined as the form which has the property that P and P^* have the same derivative when evaluated in absolutely perpendicular directions at every point of the manifold. It can be verified that P^* is uniquely defined in this way, save for sign. We may therefore speak of P^* as the form absolutely perpendicular to P. Thus P is harmonic if both P, and the form absolutely perpendicular to it, are closed.

28·2. We take the interpretation of the harmonic condition which we have just given as the basis of an approximation to the theory of harmonic integrals by means of p-sets. Let K be a covering complex of M. Then if P is any p-form, the integral $\int P$ determines a p-set $(e_p^1, \ldots, e_p^{\alpha_p})$ on K, as in § 23, which is closed if and only if P is closed. We now try to define the dual of a p-set. Since direction has no place in topology, we cannot speak of an $(n-p)$-direction perpendicular to a p-direction. But we have introduced the notion of a dual cell, and we can use the $(n-p)$-cell dual to a given p-cell in place of the direction perpendicular to a given direction. We are thus led to consider as the dual of the p-set $(e_p^1, \ldots, e_p^{\alpha_p})$ the $(n-p)$-set $(e_{n-p}^{*1}, \ldots, e_{n-p}^{*\alpha_p})$ associated with the cells of the dual complex K^*, where

$$e_{n-p}^{*i} = e_p^i \qquad (i = 1, \ldots, \alpha_p).$$

We then say that the p-set $(e_p^1, \ldots, e_p^{\alpha_p})$ is harmonic if $(e_p^1, \ldots, e_p^{\alpha_p})$ is closed in K, and if the dual set $(e_{n-p}^{*1}, \ldots, e_{n-p}^{*\alpha_p})$ is closed in K^*. The conditions for a harmonic set are therefore:

(1) $$_{(p+1)}\eta_j^i e_p^j = 0 \qquad (i = 1, \ldots, \alpha_{p+1}),$$

and (2) $$_{(n-p+1)}\eta_j^{*i} e_p^j = 0 \qquad (i = 1, \ldots, \alpha_{p-1}).$$

These equations are, however, just the equations (5) and (7) of § 23·3. The result of that paragraph therefore shows that there is exactly one harmonic p-set having arbitrarily assigned periods on R_p independent p-cycles of K.

It is tempting to make this finite representation of the harmonic property the basis of the proof of the existence of harmonic integrals with assigned periods. The details, however, of a proof of this nature present considerable difficulty, and a satisfactory proof on these lines has not yet been obtained. It may be mentioned, however, that successful solutions of some analogous problems have been obtained by these methods. Thus Courant, Friedrichs and Lewy [1] have applied this method to the problem of solving the equation

$$\frac{\partial^2 u}{\partial x^2} + \frac{\partial^2 u}{\partial y^2} = 0$$

in the (x, y) plane, where u has assigned values on the boundary of a simply-connected domain.

In my original proof [3, 4] of the existence theorem for harmonic integrals I made use of an argument based on the above considerations at one stage, but I was forced to complete the proof by other means. The proof [5] which is given below uses entirely different considerations. It is to be hoped, however, that the method which I have suggested here may yet be used to establish the existence of harmonic integrals having assigned periods, since the finite representation exhibits so clearly the essential nature of the harmonic property.

29. **Periods of harmonic integrals.** Before proceeding with the proof of the existence of a harmonic p-fold integral having assigned periods on R_p independent p-cycles of the manifold M, we must examine certain consequences of applying the bilinear relations to harmonic integrals. Let

$$P = \frac{1}{p!} P_{i_1 \ldots i_p} dx^{i_1} \ldots dx^{i_p},$$

$$Q = \frac{1}{p!} Q_{i_1 \ldots i_p} dx^{i_1} \ldots dx^{i_p}$$

be two p-forms. Then,

$$P \times Q^* = \frac{1}{p!} \sqrt{g}\, g^{i_1 j_1} \ldots g^{i_p j_p} P_{i_1 \ldots i_p} Q_{j_1 \ldots j_p} dx^1 \ldots dx^n$$

$$= Q \times P^*.$$

Also,

$$P \times P^* = \frac{1}{p!} \sqrt{g}\, g^{i_1 j_1} \ldots g^{i_p j_p} P_{i_1 \ldots i_p} P_{j_1 \ldots j_p} dx^1 \ldots dx^n.$$

Since
$$g_{ij}\,dx^i dx^j$$
is positive definite,
$$\frac{1}{p!}\sqrt{g}\,g^{i_1 j_1}\dots g^{i_p j_p}\,P_{i_1\dots i_p}P_{j_1\dots j_p} > 0$$
unless
$$P_{i_1\dots i_p} = 0.$$

Now consider the case in which the forms P and Q are harmonic and let
$$\omega^1, \dots, \omega^{R_p},$$
$$\nu^1, \dots, \nu^{R_p}$$
be the periods of the integrals
$$\int P, \quad \int Q$$
on a set of R_p independent p-cycles Γ^i_p, and let
$$\omega^{*1}, \dots, \omega^{*R_{n-p}},$$
$$\nu^{*1}, \dots, \nu^{*R_{n-p}}$$
be the periods of the duals of these integrals on a set of R_{n-p} independent $(n-p)$-cycles Γ^i_{n-p}. Let ϵ^p denote the transpose of the inverse of the intersection matrix $\| (\Gamma^i_p . \Gamma^j_{n-p}) \|$. Then
$$\epsilon^p_{ij}\,\omega^i\nu^{*j} = \int_M P \times Q^*$$
$$= \int_M Q \times P^*$$
$$= \epsilon^p_{ij}\,\nu^i\omega^{*j}$$

Also,
$$\epsilon^p_{ij}\,\omega^i\omega^{*j} = \int_M P \times P^* > 0,$$
unless $P_{i_1\dots i_p} = 0$, everywhere.

Thus a non-zero harmonic integral cannot have all its periods zero, and two distinct harmonic p-forms cannot be homologous.

30·1. **The existence theorem: preliminary considerations.** We now come to the proof of the fundamental theorem:

There exists a harmonic integral having arbitrarily assigned periods on R_p independent p-cycles of a Riemannian manifold M.

Since the harmonic form of multiplicity zero is a constant and that of multiplicity n is a constant multiple of

$$\sqrt{g}\, dx^1 \dots dx^n,$$

we may suppose the problem solved for the cases $p = 0, n$, and confine ourselves to the cases $0 < p < n$.

In the theory of potential functions in Euclidean space, we have trivial differences between the cases $n > 2$ and $n = 2$, due to the fact that if r is the distance from a point O to a variable point P the potential at P due to a unit point-charge at O is $1/r^{n-2}$ or $\log r$ according as $n > 2$ or $n = 2$. Similarly, when we consider harmonic integrals in a Riemannian manifold, we have similar differences between the cases $n > 2$ and $n = 2$. We again have to consider a function which is like $1/r^{n-2}$ when $n > 2$ and like $\log r$ when $n = 2$. We can pass from the case $n > 2$ to the case $n = 2$ by a formal change at each stage of the argument, or to avoid needless repetition we can quote the proof of the case $n = 2$, which, as we have said above, is a classical theorem on the Riemann surface. For brevity we adopt the latter course, leaving the reader to see what formal changes in our proof would be necessary in the case $n = 2$.

30·2. We shall call a function $r(x, y)$ a distance function of class v, between any two points x and y of M, if it satisfies the conditions:

(i) $r(x, y) = r(y, x) > 0, \ (x \neq y); \quad r(x, x) = 0;$

(ii) regarded as a function of x it is a function of class v of the local parameters at x;

(iii) as x approaches y we may refer the two points to the same system of local parameters. The function must then satisfy the condition

$$\lim_{x \to y} \left\{ \frac{r}{[g_{ij}(y)(x_i - y_i)(x_j - y_j)]^{\frac{1}{2}}} \right\} = 1,$$

where $g_{ij}(y)$ denotes the value of g_{ij} at the point y.

The construction of the function $r(x, y)$ is in most cases a simple matter. Let us consider first the case in which M is a manifold of class u, and can be represented as a locus of class u in a Euclidean space S_N, of finite dimensions N, given by

$$X_i = f_i(x_1, \ldots, x_n) \qquad (i = 1, \ldots, N),$$

and suppose that the metric is defined on M by the Euclidean distance element in S_N.

Then, if $P\ (x_1, \ldots, x_n)$ and $Q\ (y_1, \ldots, y_n)$ are two points of M, we may define $r(x, y)$ by the equation

$$r(x, y) = \left\{ \sum_{k=1}^{N} [f_k(x_1, \ldots, x_n) - f_k(y_1, \ldots, y_n)]^2 \right\}^{\frac{1}{2}}.$$

This applies for all values of u, including $u = \omega$, that is, the case in which M is analytic, and v may have any assigned value, $0 \leqslant v \leqslant u$. This covers a number of important applications, such as the application to algebraic varieties. Again, if the metric is not given as an intrinsic part of the definition of M, and is only to be introduced for auxiliary purposes, we may represent M as a locus of finite class $v \leqslant u$ in a Euclidean space S_N. Then we can introduce the metric by means of the Euclidean distance element in S_N, and $r(x, y)$ can be defined as above. This case is of importance in our proof of de Rham's second theorem, for which it is only necessary to define some metric, but any metric which can be defined on M will serve the purpose.

On the other hand, the metric may be given as part of the data of M and we may not be able to find a representation of

M in which the given metric can be obtained from the Euclidean metric of the surrounding space in the manner we have described. In this case we can proceed as follows. There exists a polynomial $\phi(x)$ with the properties:

(i) $$\phi'(x) > 0, \qquad 0 < a \leqslant x < a + 2b;$$

(ii) $$\phi(a) = a, \qquad \phi(a+2b) = a+b;$$

(iii) when $x = a$,

$$\frac{d}{dx}\phi(x) = 1, \quad \frac{d^r}{dx^r}\phi(x) = 0 \qquad (r = 2, \ldots, v);$$

(iv) when $x = a + 2b$,

$$\frac{d^r}{dx^r}\phi(x) = 0 \qquad (r = 1, \ldots, v).$$

Indeed, if $\psi(x)$ denotes the polynomial consisting of the first $v+1$ terms of the expansion of

$$\frac{b+x}{[2b+x]^{v+1}}$$

as a power series in x, then

$$\phi(x) = x - (x-a)^{v+1}\,\psi(x-a-2b)$$

is a polynomial with the desired properties.

Now if x is any point of M, there is a neighbourhood N_x of x with the property that any point of N_x can be joined to x by a unique geodesic lying in N_x. We can find two positive constants a, b such that the geodesic spheres Γ_x, Γ'_x, whose centres are at x and whose radii are a, $a+2b$, lie in N_x, for all points x of M. If y is any other point of M, we define $S(x, y)$ as follows:

(i) $$S(x, y) = \rho(x, y)$$

if y is within Γ_x, where $\rho(x, y)$ is the geodesic distance from y to x;

(ii) $$S(x, y) = \phi(\rho[x, y])$$

if y is within Γ'_x but not within Γ_x;

(iii) $$S(x, y) = a + b$$

otherwise.

Then $S(x, y)$ is a function of class v, whether regarded as a function of x, or of y. The function

$$r(x, y) = \tfrac{1}{2}[S(x, y) + S(y, x)]$$

satisfies our requirements. The number v can take any integral value less than u.

30·3. We now define a double form $\omega_p(x, y)$, that is a form which is a p-form in the local parameters $(x_1, \ldots, x_n)$ at x, and a p-form in the parameters $(y_1, \ldots, y_n)$ valid in the neighbourhood of y. Let $r(x, y)$ be a distance function and write

$$L = r^2.$$

Then, when $p \geqslant 0$,

$$\omega_p(x, y) = \frac{1}{(p!)^2 (-2)^p r^{n-2}} \begin{vmatrix} \dfrac{\partial^2 L}{\partial x_{i_1} \partial y_{j_1}} & \cdots & \dfrac{\partial^2 L}{\partial x_{i_1} \partial y_{j_p}} \\ \vdots & & \\ \dfrac{\partial^2 L}{\partial x_{i_p} \partial y_{j_1}} & \cdots & \dfrac{\partial^2 L}{\partial x_{i_p} \partial y_{j_p}} \end{vmatrix} dx^{i_1} \ldots dx^{i_p} dy^{j_1} \ldots dy^{j_p}.$$

When $p = -1$, we define $\omega_{-1}(x, y)$ by the equation

$$\omega_{-1}(x, y) = 0.$$

If $r(x, y)$ is of class v, the coefficients of this double form are functions of class $v-1$ of the local coordinates in the neighbourhood of any two points x, y of M, except where $x = y$. In Euclidean space, $\omega_0(x, y)$, regarded as a function of x, is the electrical potential due to a unit charge at y. In Euclidean 3-space, if we write

$$A_i dx^i = \int_\gamma \omega_1(x, y),$$

where the integration is with respect to y and is over a cycle γ, then **A** is the vector potential due to unit current in γ.

In order to use this double form to obtain a proof of the existence theorem, we have to study the behaviour of certain integrals involving $\omega_p(x,y)$ and forms derived from it, in the neighbourhood of a point O at which x and y tend to coincidence. To do this we choose local coordinates $(x_1, \ldots, x_n)$ in the neighbourhood of O, such that

(i) O is given by $x_1 = \ldots = x_n = 0$;

(ii) the fundamental quadratic form on M reduces to

$$(dx^1)^2 + \ldots + (dx^n)^2$$

at O.

We can simplify our calculations considerably by introducing an auxiliary double form, which we denote by $\bar{\omega}_p(x,y)$, defined in the neighbourhood of O. The form $\bar{\omega}_p(x,y)$ is given by the formula for $\omega_p(x,y)$, when $r(x,y)$ is replaced by

$$\rho(x,y) = [\Sigma(x_i - y_i)^2]^{\frac{1}{2}}.$$

In performing calculations involving this auxiliary double form, we shall evaluate the derived forms $\bar{\omega}_p^y(x,y)$, etc., as if the metric was $\Sigma(dx^i)^2$ everywhere.

From the definitions of $\omega_p(x,y)$ and $\bar{\omega}_p(x,y)$, it follows immediately that the orders of infinity at $x = y = 0$ of the coefficients of the forms

$$\omega_p(x,y) - \bar{\omega}_p(x,y) \quad \text{and} \quad \omega_p^y(x,y) - \bar{\omega}_p^y(x,y)$$

are $(n-3)$ and $(n-2)$ at most, respectively. If we write

$$K_p(x,y) = \omega_{p,y}^y(x,y) + (-1)^{p-1}\omega_{p-1,x}^y(x,y)$$

and $$\bar{K}_p(x,y) = \bar{\omega}_{p,y}^y(x,y) + (-1)^{p-1}\bar{\omega}_{p-1,x}^y(x,y),$$

then $K_p(x,y) - \bar{K}_p(x,y)$ is easily seen to have an infinity of order $(n-1)$, at most, at $x = y = 0$. It will be proved in § 30·4 that $\bar{K}_p(x,y)$ is identically zero, and hence it will follow that the coefficients of $r^{n-1}K(x,y)$ are finite everywhere.

Let γ be the locus

$$x_1^2+\ldots+x_n^2=a^2,$$

and δ the region $\qquad x_1^2+\ldots+x_n^2<a^2,$

where a is a sufficiently small positive constant.

From the results stated for the orders of infinity of

$$\omega_p(x,y)-\bar{\omega}_p(x,y),$$

etc., we conclude that if $u(y)$ is a regular $(p-1)$-form:

(I)

$$\lim_{a\to 0}\int_\gamma \omega_{p-1}^y(x,y)\times u(y)=\lim_{a\to 0}\int_\gamma \bar{\omega}_{p-1}^y(x,y)\times u(y);$$

(II)

$$\lim_{a\to 0}\left[\int_\delta \omega_{p-2}^y(x,y)\times u(y)\right]_x=\lim_{a\to 0}\left[\int_\delta \bar{\omega}_{p-2}^y(x,y)\times u(y)\right]_x;$$

(III)

$$\lim_{a\to 0}\left[\int_\delta u^*(y)\times\omega_{p-1}(y,x)\right]_x^x=\lim_{a\to 0}\left[\int_\delta u^*(y)\times\bar{\omega}_{p-1}(y,x)\right]_x^x;$$

(IV)

$$\lim_{a\to 0}\left[\int_\delta u^*(y)\times\omega_{p-2,y}(y,x)\right]^x=\lim_{a\to 0}\left[\int_\delta u^*(y)\times\bar{\omega}_{p-2,y}(y,x)\right]^x;$$

each expression being evaluated when x is at O.

30·4. We shall simplify our argument considerably by first proving certain results concerning the forms $\bar{\omega}_p(x,y)$ and then evaluating the right-hand members of the equalities written above. We begin by proving that $\bar{K}_p(x,y)$ is identically zero.

By direct calculation we have

$$\bar{\omega}_p^y(x,y)=\rho^{-n}(n-2)\sum(x_{i_1}-y_{i_1})\,dx^{i_2}\ldots dx^{i_{p+1}}dy^{i_{p+2}}\ldots dy^{i_n},$$

the summation being over all even derangements $(i_1, \ldots, i_n)$ of the first n integers, *each term only appearing once*. Similarly,

$$\begin{aligned}
\overline{\omega}^y_{p,y}(x,y) &= (-1)^{p-1}\rho^{-n}(n-2)(n-p)\sum dx^{i_1}\ldots dx^{i_p}dy^{i_{p+1}}\ldots dy^{i_n}\\
&\quad+(-1)^p\rho^{-n-2}n(n-2)\sum\left(\sum_{i_{p+1}}^{i_n}[x_k-y_k]^2\right)\\
&\qquad\times dx^{i_1}\ldots dx^{i_p}dy^{i_{p+1}}\ldots dy^{i_n}\\
&\quad+(-1)^{p-1}\rho^{-n-2}n(n-2)\sum\left(\sum_{i_{p+1}}^{i_n}[x_k-y_k]\,dx^k\right)\\
&\qquad\times(x_{i_1}-y_{i_1})\,dx^{i_2}\ldots dx^{i_p}dy^{i_{p+1}}\ldots dy^{i_n}\\
&=(-1)^p\rho^{-n}(n-2)\,p\sum dx^{i_1}\ldots dx^{i_p}dy^{i_{p+1}}\ldots dy^{i_n}\\
&\quad+(-1)^{p-1}\rho^{-n-2}n(n-2)\sum\left(\sum_{i_1}^{i_p}[x_k-y_k]^2\right)\\
&\qquad\times dx^{i_1}\ldots dx^{i_p}dy^{i_{p+1}}\ldots dy^{i_n}\\
&\quad+(-1)^{p-1}\rho^{-n-2}n(n-2)\sum\left(\sum_{i_{p+1}}^{i_n}[x_k-y_k]\,dx^k\right)\\
&\qquad\times(x_{i_1}-y_{i_1})\,dx^{i_2}\ldots dx^{i_p}dy^{i_{p+1}}\ldots dy^{i_n}\\
&=(-1)^p\rho^{-n}(n-2)\,p\sum dx^{i_1}\ldots dx^{i_p}dy^{i_{p+1}}\ldots dy^{i_n}\\
&\quad+(-1)^{p-1}\rho^{-n-2}n\,(n-2)\left(\sum_{k=1}^{n}[x_k-y_k]\,dx^k\right)\\
&\qquad\times\sum(x_{i_1}-y_{i_1})\,dx^{i_2}\ldots dx^{i_p}dy^{i_{p+1}}\ldots dy^{i_n}\\
&=(-1)^p\,\overline{\omega}^y_{p-1,x}(x,y).
\end{aligned}$$

Hence $$\overline{K}_p(x,y)=0.$$

30·5. We now evaluate the four integrals (I), ..., (IV), above.

(I) To evaluate

$$\lim_{a\to 0}\int_\gamma \overline{\omega}^y_{p-1}(x,y)\times u(y)$$

when x is at O, we consider first the case in which

$$u(y)=A\,dy^1\ldots dy^{p-1}.$$

Then on γ we have

$$\begin{aligned}&\overline{\omega}^y_{p-1}(x,y)\times u(y)\\ &\quad=\sum_{r=p}^{n}(-1)^{n(p-1)}a^{-n}(n-2)\,A\,dy^1\dots dy^{r-1}\,dy^{r+1}\dots dy^n\,\mu_r,\end{aligned}$$

where

$$\begin{aligned}\mu_r=(-1)^{r-1}(x_r-y_r)\,dx^1\dots dx^{p-1}&\\ +(-1)^{r-p}\sum_{i=1}^{p-1}(-1)^{i-1}(x_i-y_i)\,dx^1\dots dx^{i-1}&\\ \times dx^{i+1}\dots dx^{p-1}\,dx^r.&\end{aligned}$$

Let

$$d\Sigma=a^{-1}\sum_{k=1}^{n}(-1)^{k-1}(y_k-x_k)\,dy^1\dots dy^{k-1}\,dy^{k+1}\dots dy^n$$

be the element of volume of γ, and denote

$$a^{-n+1}\int_\gamma d\Sigma$$

by α_n. The value of α_n is independent of a. Then, since γ is a sphere, we can put

$$dy^1\dots dy^{i-1}\,dy^{i+1}\dots dy^n=(-1)^{i-1}a^{-1}(y_i-x_i)\,d\Sigma,$$

and we obtain

$$\begin{aligned}&\int_\gamma\overline{\omega}^y_{p-1}(x,y)\times u(y)\\ &\quad=(-1)^{n(p-1)-1}a^{-n-1}(n-2)\int_\gamma A\sum_p^n(y_r-x_r)^2dx^1\dots dx^{p-1}d\Sigma\\ &\qquad+(-1)^{n(p-1)-1}a^{-n-1}(n-2)\int_\gamma A\sum_{\substack{i<p\\ r\geqslant p}}(-1)^{p+i}\\ &\qquad\quad\times(y_r-x_r)(y_i-x_i)\,dx^1\dots dx^{i-1}\,dx^{i+1}\dots dx^{p-1}\,dx^r\,d\Sigma.\end{aligned}$$

Now since $u(y)$ is regular we see that the error ϵ involved by replacing A by its value A_0 at O tends to zero when a tends to zero. Therefore,

$$\int_\gamma \overline{\omega}^y_{p-1}(x,y)\times u(y)$$
$$= (-1)^{n(p-1)-1}a^{-n-1}(n-2)A_0\int_\gamma \sum_p^n (y_r-x_r)^2\,dx^1\ldots dx^{p-1}d\Sigma$$
$$+(-1)^{n(p-1)-1}a^{-n-1}(n-2)A_0\int_\gamma \sum_{\substack{i<p\\ r\geqslant p}}(-1)^{p+i}$$
$$\times (y_r-x_r)(y_i-x_i)\,dx^1\ldots dx^{i-1}dx^{i+1}\ldots dx^{p-1}dx^r\,d\Sigma+\epsilon$$
$$= (-1)^{n(p-1)-1}(n-2)A_0 n^{-1}(n-p+1)\alpha_n\,dx^1\ldots dx^{p-1}+0+\epsilon$$
$$= (-1)^{n(p-1)-1}n^{-1}(n-2)(n-p+1)\alpha_n[u(x)]_{x=0}+\epsilon,$$

where ϵ tends to zero with a.

Hence we deduce that for any regular $(p-1)$-form u,

$$\lim_{a\to 0}\int_\gamma \overline{\omega}^y_{p-1}(x,y)\times u(y)$$
$$= (-1)^{n(p-1)-1}n^{-1}(n-2)(n-p+1)\alpha_n u(x).$$

When we consider the other limits listed above, it is sufficient to consider the case

$$u(y) = A\,dy^1\ldots dy^{p-1},$$

as in the calculation just given. To calculate the second limit, we have

(II) $\overline{\omega}^y_{p-2}(x,y)\times u(y)$

$$= (-1)^{(n-1)(p-1)}(n-2)\,dy^1\ldots dy^n$$
$$\times A\sum_1^{p-1}(-1)^{i-1}\rho^{-n}(x_i-y_i)\,dx^1\ldots dx^{i-1}dx^{i+1}\ldots dx^{p-1}$$
$$= (-1)^{(n-1)(p-1)}dy^1\ldots dy^n$$
$$\times A\sum_1^{p-1}(-1)^{i-1}\frac{\partial}{\partial y_i}\left(\frac{1}{\rho^{n-2}}\right)dx^1\ldots dx^{i-1}dx^{i+1}\ldots dx^{p-1}$$
$$= (-1)^{(n-1)(p-1)}dy^1\ldots dy^n$$
$$\times\sum_1^{p-1}(-1)^{i-1}\left\{\frac{\partial}{\partial y_i}\left(\frac{A}{\rho^{n-2}}\right)-\frac{1}{\rho^{n-2}}\frac{\partial A}{\partial y_i}\right\}$$
$$\times\,dx^1\ldots dx^{i-1}dx^{i+1}\ldots dx^{p-1}.$$

Hence

$$\int_\delta \overline{\omega}^y_{p-2}(x,y) \times u(y)$$

$$= (-1)^{(n-1)(p-1)} \int_\gamma A\rho^{-n+2} \sum_1^{p-1} dx^1 \dots dx^{i-1}$$

$$\times dx^{i+1} \dots dx^{p-1} dy^1 \dots dy^{i-1} dy^{i+1} \dots dy^n$$

$$-(-1)^{(n-1)(p-1)} \int_\delta \rho^{-n+2} \sum_1^{p-1} (-1)^{i-1} \frac{\partial A}{\partial y_i}$$

$$\times dx^1 \dots dx^{i-1} dx^{i+1} \dots dx^{p-1} dy^1 \dots dy^n.$$

Hence

$$\left[\int_\delta \overline{\omega}^y_{p-2}(x,y) \times u(y)\right]_x$$

$$= (-1)^{(n-1)(p-1)} (n-2) \int_\gamma a^{-n-1} A \left[\sum_1^{p-1} (y_k - x_k)^2 dx^1 \dots dx^{p-1}\right.$$

$$+ \sum_{\substack{i<p \\ j \geqslant p}} (-1)^{i-1} (y_i - x_i)(y_j - x_j) dx^j dx^1 \dots dx^{i-1}$$

$$\left. \times dx^{i+1} \dots dx^{p-1}\right] d\Sigma + \epsilon$$

$$= (-1)^{(n-1)(p-1)} n^{-1}(n-2)(p-1) \alpha_n A_0 dx^1 \dots dx^{p-1} + \epsilon,$$

where the terms included in ϵ tend to zero with a. Hence, for any regular u,

$$\lim_{a \to 0} \left[\int_\delta \overline{\omega}^y_{p-2}(x,y) \times u(y)\right]_x$$

$$= (-1)^{(n-1)(p-1)} n^{-1}(n-2)(p-1) \alpha_n u(x).$$

(III) We now evaluate

$$\lim_{a \to 0} \left[\int_\delta u^*(y) \times \overline{\omega}_{p-1}(y,x)\right]^x_x.$$

We consider $\quad u^*(y) = A\, dy^p \dots dy^n.$

Then,

$$\left[\int_\delta u^*(y)\times\overline{\omega}_{p-1}(y,x)\right]_x$$

$$= (-1)^{(n-1)(p-1)}\left[\int_\delta \rho^{-n+2} A\,dx^1\dots dx^{p-1}\,dy^1\dots dy^n\right]_x$$

$$= (-1)^{n(p-1)-1}\left[\int_\delta \sum_{k=p}^{n}\frac{\partial}{\partial y_k}\left(\frac{A}{\rho^{n-2}}\right)dx^1\dots dx^{p-1}\,dx^k\,dy^1\dots dy^n\right.$$

$$\left.-\int_\delta \rho^{-n+2}\sum_{k=p}^{n}\frac{\partial}{\partial y_k}A\,dx^1\dots dx^{p-1}\,dx^k\,dy^1\dots dy^n\right]$$

$$= (-1)^{n(p-1)-1}\left[\int_\gamma \sum_p^n(-1)^{k-1}\rho^{-n+2}A\,dx^1\dots dx^{p-1}\,dx^k\right.$$

$$\left.\times dy^1\dots dy^{k-1}\,dy^{k+1}\dots dy^n\right]$$

$$+(-1)^{n(p-1)}\left[\int_\delta \rho^{-n+2}\sum_p^n\frac{\partial A}{\partial y_k}dx^1\dots dx^{p-1}\,dx^k\,dy^1\dots dy^n\right],$$

since
$$\frac{\partial}{\partial x_k}\left(\frac{1}{\rho^{n-2}}\right) = -\frac{\partial}{\partial y_k}\left(\frac{1}{\rho^{n-2}}\right).$$

Now
$$\left[\int_\gamma \sum_p^n(-1)^{k-1}\rho^{-n+2}A\,dx^1\dots dx^{p-1}\,dx^k\right.$$

$$\left.\times dy^1\dots dy^{k-1}\,dy^{k+1}\dots dy^n\right]^*_x$$

$$=(-1)^{p-1}\left[\int_\gamma \sum_{k=p}^{n}\rho^{-n+2}A\,dx^p\dots dx^{k-1}\,dx^{k+1}\dots dx^n\right.$$

$$\left.\times dy^1\dots dy^{k-1}\,dy^{k+1}\dots dy^n\right]_x$$

$$= (-1)^{p-1}(n-2)\left[\int_\gamma \rho^{-n}A\sum_{k=p}^{n}\left(\sum_{j=1}^{n}[y_j-x_j]\,dx^j\right)\right.$$

$$\left.\times dx^p\dots dx^{k-1}\,dx^{k+1}\dots dx^n\,dy^1\dots dy^{k-1}\,dy^{k+1}\dots dy^n\right]$$

$$= (n-2)\int_\gamma \rho^{-n-1}A\sum_p^n(y_k-x_k)^2\,dx^p\dots dx^n\,d\Sigma$$

$$+(-1)^{p-1}(n-2)\int_\gamma \sum_{\substack{j<p\\ k\geqslant p}}\rho^{-n-1}A(y_j-x_j)(y_k-x_k)$$

$$\times dx^j\,dx^p\dots dx^{k-1}\,dx^{k+1}\dots dx^n\,d\Sigma$$

$$= n^{-1}(n-2)(n-p+1)\,\alpha_n A_0\,dx^p\dots dx^n+\epsilon,$$

where ϵ tends to zero with a, and A_0 is the value of A at O, by reasoning similar to that given before. It is easily shown that

$$\lim_{a\to 0}\left[\int_\delta \rho^{-n+2}\frac{\partial A}{\partial y_k}dx^1\ldots dx^{p-1}dx^k dy^1\ldots dy^n\right]^*_x = 0.$$

Hence we conclude that for any regular u

$$\lim_{a\to 0}\left[\int_\delta u^*(y)\times\overline{\omega}_{p-1}(y,x)\right]^x_x$$

$$= (-1)^{n(p-1)-1}n^{-1}(n-2)(n-p+1)\alpha_n u(x).$$

(IV) Finally, we have

$$\int_\delta u^*(y)\times\overline{\omega}_{p-2,y}(y,x)$$

$$= (-1)^{n-p+1}\int_\gamma u^*(y)\times\overline{\omega}_{p-2}(y,x)$$

$$+(-1)^{n-p}\int_\delta u^*_y(y)\times\overline{\omega}_{p-2}(y,x).$$

Hence

$$\left[\int_\delta u^*(y)\times\overline{\omega}_{p-2,y}(y,x)\right]^x = (-1)^{n-p+1}\int_\gamma u^*(y)\times\overline{\omega}^x_{p-2}(y,x)+\epsilon.$$

By a calculation similar to one performed above, we see that the right-hand member reduces to

$$(-1)^{(n-1)(p-1)}n^{-1}(n-2)(p-1)\alpha_n u(x),$$

when a tends to zero.

31·1. The existence theorem, continued. We now show how the calculations of the previous paragraph can be used to reduce the proof of the existence theorem to the solution of a certain integral equation, and obtain some results which are of importance in the solution of this equation.

The first step towards constructing a harmonic integral with assigned periods is to construct a regular closed form A whose integral has the given periods. We saw how to do this in Chapter II. Another regular closed form whose integral has the given periods is given by

$$A - u_x,$$

where u is a $(p-1)$-form. Our problem is to find a form u such that this is harmonic, that is, we have to solve the equation in u,

$$u^x{}_x = A^*{}_x.$$

We have therefore to consider the problem of solving the equation

$$u^x{}_x = f, \tag{1}$$

where $$f \sim 0.$$

Let $\varDelta$ denote an n-dimensional chain of M having boundary $\varGamma$, and let u and v be two $(p-1)$-forms which are regular in a domain containing $\varDelta$ and $\varGamma$ in its interior. From the equations

$$(u^x \times v)_x = u^x{}_x \times v + (-1)^{n-p}\, u^x \times v_x,$$

$$(v^x \times u)_x = v^x{}_x \times u + (-1)^{n-p}\, v^x \times u_x,$$

$$u^x \times v_x = v^x \times u_x,$$

we obtain the equation

$$\int_{\varDelta} u^x{}_x \times v - \int_{\varDelta} v^x{}_x \times u = \int_{\varGamma} u^x \times v - \int_{\varGamma} v^x \times u. \tag{2}$$

Let δ and γ be defined as in the previous paragraph, and let

$$\varDelta = M - \delta,$$

$$\varGamma = -\gamma.$$

In equation (2) we take $v = \omega_{p-1}(x, y)$. Remembering that when a, the radius of γ, tends to zero we can replace $\omega_{p-1}(x, y)$ by $\overline{\omega}_{p-1}(x, y)$ when integrating over γ or δ, we have

$$\int_{\varDelta} \omega^y_{p-1,\,y}(x, y) \times u(y) = \int_{\varDelta} u^y{}_y \times \omega_{p-1}(x, y)$$
$$+ (-1)^{n(p-1)}\, n^{-1}(n-2)\,(n-p+1)\,\alpha_n u(x) + \xi,$$

where ξ tends to zero with a, by (I) of § 30·5. Also,

$$\begin{aligned}\int_{\Delta} \omega^{y}_{p-2,x}(x,y) \times u(y) &= \left[\int_{M} \omega^{y}_{p-2}(x,y) \times u(y)\right]_{x} - \left[\int_{\delta} \omega^{y}_{p-2}(x,y) \times u(y)\right]_{x} \\ &= \left[\int_{M} \omega^{y}_{p-2}(x,y) \times u(y)\right]_{x} \\ &\qquad - (-1)^{(n-1)(p-1)} n^{-1}(n-2)(p-1)\,\alpha_n u(x) + \eta,\end{aligned}$$

by (II) of § 30·5, where η tends to zero with a. Combining these two results, we have

$$\begin{aligned}\int_{M} K_{p-1}(x,y) \times u(y) - (-1)^{n(p-1)}(n-2)\,\alpha_n u(x) &- \int_{M} u^{y}{}_{y}(y) \times \omega_{p-1}(y,x) \\ &= (-1)^{p}\left[\int_{M} \omega^{y}_{p-2}(x,y) \times u(y)\right]_{x} + \zeta \\ &= (-1)^{n}\int_{M} u^{*}_{y}(y) \times \omega_{p-2,x}(y,x) + \zeta, \qquad (3)\end{aligned}$$

where ζ tends to zero with a. Since the other terms in the equation are independent of a, $\zeta = 0$. Similarly we have, using (III) and (IV) of § 30·5,

$$\begin{aligned}\int_{M} u^{*}(y) \times K_{p-1}(y,x) &= \int_{\Delta} u^{*}(y) \times K_{p-1}(y,x) + \int_{\delta} u^{*}(y) \times K_{p-1}(y,x) \\ &= \left[\int_{\Delta} u^{*}(y) \times \omega_{p-1}(y,x)\right]^{x}_{x} + (-1)^{p}\left[\int_{\Delta} u^{*}(y) \times \omega_{p-2,y}(y,x)\right]^{x} \\ &\qquad + \int_{\delta} u^{*}(y) \times K_{p-1}(y,x) \\ &= \left[\int_{M} u^{*}(y) \times \omega_{p-1}(y,x)\right]^{x}_{x} + (-1)^{p}\left[\int_{M} u^{*}(y) \times \omega_{p-2,y}(y,x)\right]^{x} \\ &\qquad + (-1)^{n(p-1)} n^{-1}(n-2)(n-p+1)\,\alpha_n u^{*}(x) \\ &\qquad + (-1)^{n(p-1)} n^{-1}(n-2)(p-1)\,\alpha_n u^{*}(x) + \xi,\end{aligned}$$

where ξ tends to zero with a. Therefore, by a simple calculation, we have

$$\left[\int_M u^*(y) \times \omega_{p-1}(y, x)\right]_x^x$$

$$= \int_M u^*(y) \times K_{p-1}(y, x) - (-1)^{n(p-1)} (n-2)\, \alpha_n u^*(x)$$

$$- (-1)^n \int_M u_y^*(y) \times \omega_{p-2}^x(y, x). \qquad (4)$$

31·2. From equation (3) we observe that if $u(x)$ is a closed form, then

$$\int_M K_{p-1}(x, y) \times u(y)$$

is also a closed form, and that if $u(x)$ is null, then

$$\int_M K_{p-1}(x, y) \times u(y)$$

is also null. Of more importance in what follows is the corresponding result which can be deduced from equation (4). From this equation we see that

$$\int_M u^*(y) \times K_{p-1}(y, x)$$

is closed when $u^*(x)$ is closed, and null when $u^*(x)$ is null.

31·3. From equation (3) we see that, if u is any solution of (1), u satisfies

$$\int_M K_{p-1}(x, y) \times u(y) - (-1)^{n(p-1)} (n-2)\, \alpha_n u(x)$$

$$= \int_M f(y) \times \omega_{p-1}(y, x) + (-1)^n \int_M u_y^*(y) \times \omega_{p-2,\,x}(y, x).$$

We notice, however, that if u is any solution of (1), so is $u + v_x$, where v is a $(p-2)$-form. Now, if we suppose that we can solve equation (1) when p is replaced by $p-1$, and the unknown form is a $(p-2)$-form and the second member is a null $(n-p+2)$-form, we can choose v so that

$$v^x{}_x = -u^*{}_x.$$

Replacing u by $u+v_x$, we can therefore suppose that

$$u^* \to 0,$$

and u is then a solution of the integral equation

$$\int_M K_{p-1}(x,y) \times u(y) - (-1)^{n(p-1)}(n-2)\,\alpha_n u(x)$$

$$= \int_M f(y) \times \omega_{p-1}(y,x). \qquad (5)$$

We shall proceed to solve our existence problem by considering the solution of this integral equation. The method will be one of induction on p; and in the course of the proof of the existence of a harmonic p-fold integral with assigned periods we shall show that equation (1) has a solution whenever f is a null $(n-p+2)$-form. Consequently, it is permissible to suppose that

$$u^* \to 0.$$

At the same time it should be observed that this assumption is not really necessary; we merely solve (5) and deduce from the solution obtained a solution of (1), and the above remark is made merely to show that the omission of the term

$$(-1)^n \int_M u^*_y(y) \times \omega_{p-2,\,x}(y,x)$$

from (3) has an obvious explanation.

32·1. Digression on the solution of integral equations. Equation (5) is equivalent to $N = \binom{n}{p-1}$ equations in the coefficients of the form u. If we write these coefficients in some order as

$$u_1, \ldots, u_N,$$

the equations are of the form

$$u_i(x) + \lambda \int_M K_i^j(x, y)\, u_j(y)\, d\tau_y = f_i(x) \qquad (i = 1, \ldots, N), \quad \text{(A)}$$

where the functions $K_i^j(x, y)$ are continuous in x or y except at $x = y$, and in applying the summation convention we sum from 1 to N. The parameter λ is introduced for convenience, and $d\tau_y$ is an element of volume, the suffix y indicating that the integration is with respect to the variables $(y_1, \ldots, y_n)$. The equations (A) form a simultaneous set of Fredholm equations of the second kind, and the reader is referred to the standard works on integral equations for the solution (see, for instance, Hilbert [2] or Kowalewski [6]). In the present case $K_i^j(x, y)$ becomes infinite when $x = y$, but, since the coefficients of $[r(x, y)]^{n-1} K_i^j(x, y)$ are finite everywhere, the necessary and sufficient conditions that (A) should have a solution are of the same form as if $K_i^j(x, y)$ was finite everywhere.

The following are the main facts which we require to know concerning the solution of equations (A).

I. If the parameter λ has any value which is not a zero of a certain integral function $D(\lambda)$, there exists one and only one set of continuous functions $u_i(x)$ which satisfy (A). The functions are all zero if $f_i(x)$ is zero for each i. The solutions can be written as infinite series in the form

$$D(\lambda)\, u_i(x) = D(\lambda) f_i(x) + \sum_1^\infty \lambda^r \int_M L_{(r)i}^j(x, y)\, f_j(y)\, d\tau_y.$$

The series on the right converges for all finite values of λ.

II. If $\lambda = \lambda_1$ is a zero of $D(\lambda)$, the set of homogeneous equations

$$u_i(x) + \lambda_1 \int_M K_i^j(x, y)\, u_j(y)\, d\tau_y = 0 \qquad (i = 1, \ldots, N) \quad \text{(B)}$$

has a positive finite number of linearly independent sets of solutions, and moreover the associated set of homogeneous equations

$$v^{*i}(x) + \lambda_1 \int_M v^{*j}(y)\, K^i_j(y, x)\, d\tau_y = 0 \qquad (i = 1, \ldots, N) \quad \text{(C)}$$

has the same number of linearly independent sets of solutions. The zeros of $D(\lambda)$ are called the *characteristic values* of the parameter in (A). If λ_1 is a characteristic value of the parameter, equation (A), with $\lambda = \lambda_1$, has a solution if and only if

$$\int_M v^{*i}(x) f_i(x)\, d\tau_x = 0$$

for each set $v^{*i}(x)$ of solutions of the associated equation (C). Moreover, if the conditions are satisfied by the functions $f_i(x)$, and if

$$u_1(x, \lambda), \ \ldots, \ u_N(x, \lambda)$$

is the set of solutions of (A) for a general value of the parameter near λ_1, then the limits

$$\lim_{\lambda \to \lambda_1} u_1(x, \lambda), \ \ldots, \ \lim_{\lambda \to \lambda_1} u_N(x, \lambda)$$

exist, and give a set of solutions of (A) with $\lambda = \lambda_1$; and the most general solution is obtained by adding an arbitrary solution of (B).

III. The non-homogeneous equations

$$v^{*i}(x) + \lambda \int_M v^j(y)\, K^i_j(y, x)\, d\tau_y = g^i(x) \qquad (i = 1, \ldots, N) \quad \text{(D)}$$

have the same characteristic values as the equation (A). If $D(\lambda) \neq 0$, the solution of (D) is given by

$$D(\lambda)\, [v^i(x) - g^i(x)] = \sum_1^\infty \lambda^r \int_M g^j(y)\, L^i_{(r)j}(y, x)\, d\tau_y,$$

and if $D(\lambda) = 0$, the equation (D) has a solution if and only if

$$\int_M u_i(x)\, g^i(x)\, d\tau_x = 0,$$

for each set of solutions $u_i(x)$ of (B). The other properties, stated in (II), of the equation (A) can now be applied to (D).

IV. In the present case $[r(x,y)]^{n-1} K_i^j(x,y)$ is a function of class $(v-3)$ of the local coordinates, when regarded as a function of x or y. In this case we can say that the solutions of (B) are functions of class $(v-3)$ of the local parameters.

32·2. When we express these results in the notation of forms, we obtain theorems which will enable us to solve equations (1). Along with the equation

$$\int_M K_{p-1}(x,y) \times u(y) - (-1)^{n(p-1)} (n-2)\, \alpha_n\, u(x) = \int_M f(y) \times \omega_{p-1}(y,x), \qquad (5)$$

we consider the homogeneous equation

$$\int_M K_{p-1}(x,y) \times u(y) - (-1)^{n(p-1)} (n-2)\, \alpha_n\, u(x) = 0 \qquad (6)$$

and also the associated equation

$$\int_M v^*(y) \times K_{p-1}(y,x) - (-1)^{n(p-1)} (n-2)\, \alpha_n\, v^*(x) = 0, \qquad (7)$$

where $v(x)$ is a $(p-1)$-form. Equations (6) and (7) have the same number m of linearly independent solutions. If $m = 0$, (5) has a unique solution, whatever the form $f(y)$ may be. If $m > 0$, (5) has a solution if and only if

$$\int_M \int_M f(y) \times \omega_{p-1}(y,x) \times v^*(x) = 0, \qquad (8)$$

for each solution $v^*(x)$ of (7).

32·3. We now prove a lemma which will be used later in this chapter.

Let us consider the non-homogeneous equation

$$\lambda \int_M u^*(y) \times K_{p-1}(y,x) - (-1)^{n(p-1)} (n-2)\, \alpha_n\, u^*(x) = F(x), \qquad (7)'$$

where $K_{p-1}(x, y)$ is the form defined in § 30·3. We are only interested in values of λ near $\lambda = 1$, and we may suppose that λ is restricted to a domain containing no characteristic value of λ different from $\lambda = 1$, which may or may not be characteristic. If $\lambda = 1$ is a characteristic value, we assume that the $(n-p+1)$-form $F(x)$ satisfies the conditions for the existence of a solution of (7)′ when $\lambda = 1$.

The unique solution of (7)′ for a general value of λ is given by

$$D(\lambda)\,[u^*(x) - F(x)] = \sum_1^\infty \lambda^n \int_M F(y) \times L_{(n)}(y, x).$$

The series on the right converges for all finite values of λ, and the integral function $D(\lambda)$ does not vanish for any value of λ in the domain, other than $\lambda = 1$. The function vanishes for $\lambda = 1$ if and only if (6) and (7) have non-zero solutions.

We saw in § 31·2 that if $u^*(x)$ is closed, then

$$\int_M u^*(y) \times K_{p-1}(y, x)$$

is also closed. From the manner in which the forms $L_{(n)}(y, x)$ are constructed, it can easily be shown that this implies that

$$\int u^*(y) \times L_{(n)}(y, x)$$

is closed. Hence if $F(x)$ is closed, the solution of (7)′ is closed.

Now let λ tend to 1. The solution of (7)′ tends to a solution of the equation in which λ is replaced by 1, and it is easily shown that this solution is closed.

We can now make the following deduction. Let $u^*(x)$ be any regular form such that

$$F(x) = \int_M u^*(y) \times K_{p-1}(y, x) - (-1)^{n(p-1)}\,(n-2)\,\alpha_n\, u^*(x) \to 0.$$

There exists a solution $v^*(x)$ of (7)′ (with λ replaced by 1) which is closed. Hence $u^*(x) - v^*(x)$ is either zero or else a solution of (7).

Since, by § 31·2, if $u^*(x)$ is a null form,

$$\int_M u^*(y) \times K_{p-1}(y, x) \sim 0,$$

we can, by a similar argument, show that if $u^*(x)$ is any regular form such that

$$\int_M u^*(y) \times K_{p-1}(y, x) - (-1)^{n(p-1)} (n-2)\, \alpha_n\, u^*(x) \sim 0,$$

there exists a null form $v^*(x)$, such that $u^*(x) - v^*(x)$ is either zero or a solution of (7). Finally, on account of the last results of § 32·1, we can say that any solution of the homogeneous equation (6) is a form whose coefficients are functions of class $(v-3)$ of the local parameters.

33·1. **The existence theorem, concluded.** Since there does not exist a harmonic integral with all its periods zero, the number of harmonic integrals of multiplicity p which are linearly independent cannot exceed R_p. But $R_0 = 1$, and we have seen that a constant c is a harmonic integral of multiplicity zero. Therefore the existence theorem is proved in the case $p = 0$. We may use induction to establish the result in the general case, and for this purpose we shall assume the truth of the theorem for $(p-1)$-fold integrals.

Let us denote by

$$u_{p-1}^i \qquad (i = 1, \ldots, R_{p-1})$$

the harmonic forms of multiplicity $(p-1)$. Substitute u_{p-1}^i for u in (3). Since

$$u_{p-1}^i \to 0, \qquad [u_{p-1}^i]^* \to 0,$$

we see that u_{p-1}^i is a solution of the homogeneous equation (6). Consider any other solution of (6) which is closed. Since there are R_{p-1} independent harmonic $(p-1)$-forms on M, any closed $(p-1)$-form is equal to the sum of a harmonic form and a form whose integral has zero periods. We may therefore take the set of closed forms which satisfy (6) to be the harmonic forms and the forms

$$\phi^i \qquad (i=1, \dots, S_{p-1}),$$

whose integrals have zero periods. Finally, suppose that there are T_{p-1} solutions of (6), no linear combination of which is closed, and denote them by

$$\psi^i \qquad (i=1, \dots, T_{p-1}).$$

If we substitute for ϕ^i and ψ^i for $u(y)$ in (3), we obtain the equations

$$\int_M [\phi^i(y)]^*{}_y \times \omega_{p-2}(y,x) \to 0,$$

and

$$\int_M [\psi^i(y)]^v{}_y \times \omega_{p-1}(y,x) \sim 0,$$

and from (4) we conclude that

$$\int_M [\phi^i(y)]^* \times K_{p-1}(y,x) - (-1)^{n(p-1)}(n-2)\,\alpha_n[\phi^i(x)]^* = \left[\int_M [\phi^i(y)]^* \times \omega_{p-1}(y,x)\right]^x_x, \quad (9)$$

and

$$\int_M [\psi^i(y)]^v{}_y \times K_{p-1}(y,x) - (-1)^{n(p-1)}(n-2)\,\alpha_n[\psi^i(x)]^x{}_x = 0. \quad (10)$$

33·2. From the theory of integral equations, we know that the associated homogeneous equation (7) has $R_{p-1}+S_{p-1}+T_{p-1}$ independent solutions. From among these we can write down T_{p-1} at once. For, equation (10) tells us that

$$\psi^{ix}{}_x \qquad (i=1, \dots, T_{p-1})$$

are solutions. Further they are independent; indeed if

$$\Sigma a_i \psi^{ix}{}_x = 0,$$

then

$$\int a_i \psi^i_x$$

would be a harmonic integral without periods and hence it would be zero. Therefore

$$a_i \psi^i \to 0,$$

which is contrary to our hypothesis. Now consider (9), and use the lemma of § 32·3. We have solutions

$$\phi^{i*} - \xi^i \qquad (i = 1, \ldots, S_{p-1}),$$

of (7), where ξ^i is a null form. If these solutions are connected by a relation

$$a_i(\phi^{i*} - \xi^i) \to 0,$$

then
$$a_i \phi^{i*} \to 0,$$

and hence
$$\int a_i \phi^i$$

would be a harmonic integral without periods, which is not possible. The $S_{p-1} + T_{p-1}$ solutions of (7) which we have found are therefore independent. Again, if we substitute u^{i*}_{p-1} for u^* in (4) we obtain the equation

$$\int_M [u^i_{p-1}(y)]^* \times K_{p-1}(y,x) - (-1)^{n(p-1)}(n-2)\,\alpha_n [u^i_{p-1}(x)]^*$$
$$= \left[\int_M [u^i_{p-1}(y)]^* \times \omega_{p-1}(y,x)\right]^x_x$$

and by an argument similar to that given above we deduce the existence of R_{p-1} further solutions

$$u^{i*}_{p-1} - \eta^i, \quad \eta^i \sim 0 \qquad (i = 1, \ldots, R_{p-1})$$

of (7), and show that these, with the $S_{p-1} + T_{p-1}$ solutions already found, give a complete set of independent solutions of (7). It further follows that any solution of (7) which is null can be written in the form

$$a_i \psi^{ix}{}_x.$$

33·3. Before applying the test (8) for the solvability of (5), it is convenient at this stage to prove two lemmas.

I. *If v^* is any solution of* (7), *then*

$$u = \int_M v^*(y) \times \omega_{p-1}(y, x) \to 0.$$

From equation (4), and the fact that v^* is a solution of (7), we have

$$u^x{}_x = (-1)^{n-1}\left[\left[\int_M v^*_y(y) \times \omega_{p-2}(y, x)\right]\right]^x,$$

and hence $$u^{xx} \sim 0.$$

Therefore $$\int u^x{}_x$$

is a harmonic integral with all its periods zero, and so

$$u^x{}_x = 0.$$

Therefore $$\int u_x$$

is a harmonic integral, and all its periods are zero; hence

$$u \to 0.$$

II. *The most general solution v^* of*

$$\int v^*(y) \times \omega_{p-1}(y, x) \to 0$$

which is null is given by

$$v^* = a_i \psi^{ix}{}_x.$$

By (4) it follows that v^* is a null form which satisfies (7), and hence

$$v^* = a_i \psi^{ix}{}_x.$$

33·4. Equation (5) will have a solution if and only if

$$\int_M \int_M f(y) \times \omega_{p-1}(y, x) \times v^*(x) = 0$$

for each solution v^* of (7). By Lemma I we know that

$$\int_M \omega_{p-1}(y, x) \times v^*(x) \to 0,$$

and the condition is therefore satisfied, since

$$f \sim 0.$$

Let u be a solution. From (3), we have

$$\int_M [u^y_y(y) - f(y)] \times \omega_{p-1}(y, x) \sim 0,$$

and
$$u^x{}_x - f \sim 0.$$

Hence, by Lemma II, we have

$$u^x{}_x - f = \Sigma a_i \psi^{ix}{}_x,$$

where the coefficients a_i are constants. Hence, if

$$U = u - a_i \psi^i,$$

$$U^x{}_x = f.$$

Therefore
$$\int (A - U_x)$$

is a harmonic integral with the assigned periods. In order that this harmonic form should be regular U must be of class $v - 3 \geqslant 3$, i.e. $v \geqslant 6$.

We have thus established our existence theorem for a general value of p. We have seen in the course of the proof that a harmonic p-form satisfies the equation

$$\int_M K_p(x, y) \times u(y) - (-1)^{np} (n-2) \alpha_n u(x) = 0.$$

If we now refer to the results quoted in § 32·2 regarding integral equations, we conclude that the coefficients of a harmonic form are functions of class $v - 3$ of the local parameters. If $v = \omega$, that is, if the manifold is analytic, and it is possible to introduce an analytic distance function, the coefficients of a harmonic form are analytic.

34. **De Rham's second theorem.** It should be observed that in the proof of the existence of a p-fold harmonic integral, we did not make any use of the theorem that a closed form

whose integral has all its periods zero is a null form. We are now in a position to deduce this theorem.

On any manifold M of class $u \geqslant 7$ we can, as we have seen, introduce a metric of finite class $v = 6$. Suppose this is done, and let

$$\int \phi$$

be the integral of a closed form which has all its periods zero. Then, by the result of § 33·4, there exists a closed form U such that

$$\int (\phi - U_x)$$

is a harmonic integral. But since this integral has no periods,

$$\phi - U_x = 0,$$

and hence ϕ is a null form.

35. The equations satisfied by a harmonic tensor.

If $\int P$ is a harmonic integral, where

$$P = \frac{1}{p!} P_{i_1 \ldots i_p} dx^{i_1} \ldots dx^{i_p},$$

we say that the tensor $P_{i_1, \ldots, i_p}$ is *a harmonic tensor*. Let us now consider the differential equations satisfied by a harmonic tensor. The condition

$$P \rightarrow 0$$

is equivalent to

$$\sum_{r=1}^{p+1} (-1)^{r-1} \frac{\partial}{\partial x_{i_r}} P_{i_1 \ldots i_{r-1} i_{r+1} \ldots i_{p+1}} = 0,$$

in any coordinate system. Consider, in particular, a coordinate system $(y_1, \ldots, y_n)$, geodesic at a point O. At O we have

$$\frac{\partial}{\partial y_k} P_{i_1 \ldots i_p} = P_{i_1 \ldots i_p, k}$$

and hence the condition at O is given by

$$\sum_{r=0}^{p+1}(-1)^{r-1}P_{i_1\dots i_{r-1}i_{r+1}\dots i_p,\, i_r}=0,$$

that is, $$\delta^{j_1\dots j_{p+1}}_{i_1\dots i_{p+1}}P_{j_1\dots j_p,\, j_{p+1}}=0.$$

Hence in any coordinate system

$$\delta^{j_1\dots\ p+1}_{i_1\dots i_{p+1}}P_{j_1\dots j_p,\, j_{p+1}}=0. \tag{11}$$

Conversely if (11) is satisfied at all points of M, we have, at any point O,

$$\sum_{r=1}^{p+1}(-1)^{r-1}\frac{\partial}{\partial y_{i_r}}P_{i_1\dots i_{r-1}i_{r+1}\dots i_{p+1}}=0,$$

in geodesic coordinates at O, and hence in all coordinate systems. Hence (11) is the necessary and sufficient condition that

$$P\to 0.$$

Again, we have similarly as the condition that P^* should be closed,

$$\delta^{j_1\dots j_{n-p+1}}_{i_1\dots i_{n-p+1}}P^*_{j_1\dots j_{n-p},\, j_{n-p+1}}=0,$$

that is,

$$\delta^{j_1\dots j_{n-p+1}}_{i_1\dots i_{n-p+1}}[\sqrt{g}\,\epsilon_{k_1\dots k_p j_1\dots j_{n-p}}g^{k_1l_1}\dots g^{k_pl_p}P_{l_1\dots l_p}],_{\,j_{n-p+1}}=0,$$

that is,

$$\sqrt{g}\,\epsilon_{k_1\dots k_p j_1\dots j_{n-p}}\delta^{j_1\dots j_{n-p+1}}_{i_1\dots i_{n-p+1}}g^{k_1l_1}\dots g^{k_pl_p}P_{l_1\dots l_p,\, j_{n-p+1}}=0.$$

Since (g^{ij}) is a non-singular matrix, we may replace this last equation by the equation obtained by multiplying it by

$$\sqrt{g}\,g^{a_1i_1}\dots g^{a_{n-p+1}i_{n-p+1}}.$$

We obtain the equation

$$g\begin{vmatrix} g^{a_1j_1} & \dots & g^{a_1j_{n-p+1}} \\ \vdots & & \\ g^{a_{n-p+1}j_1} & \dots & g^{a_{n-p+1}j_{n-p+1}} \end{vmatrix}\epsilon_{k_1\dots k_p j_1\dots j_{n-p}}g^{k_1l_1}\dots g^{k_pl_p}P_{l_1\dots l_p,\, j_{n-p+1}}=0.$$

We now expand the determinant in terms of the last column, and use Jacobi's theorem that

$$g\begin{vmatrix} g^{r_1 s_1} & \dots & g^{r_1 s_{n-p}} \\ \vdots & & \\ g^{r_{n-p} s_1} & \dots & g^{r_{n-p} s_{n-p}} \end{vmatrix} = \begin{vmatrix} g_{r_{n-p+1} s_{n-p+1}} & \dots & g_{r_{n-p+1} s_n} \\ \vdots & & \\ g_{r_n s_{n-p+1}} & \dots & g_{r_n s_n} \end{vmatrix},$$

where $(r_1, \dots, r_n)$ and $(s_1, \dots, s_n)$ are like derangements of the first n integers. We have, save for a non-zero numerical factor,

$$g^{rs}\,\delta^{l_1 \dots\dots\dots\dots l_p}_{r\, a_{n-p+2} \dots a_n} P_{l_1 \dots l_p, s} = 0,$$

i.e.

$$g^{rs} P_{i_1 \dots i_{p-1} r, s} = 0. \tag{12}$$

This equation could also have been obtained by using geodesic coordinates. The conditions that $P_{i_1, \dots, i_p}$ is a harmonic integral can therefore be taken as (11) and (12).

REFERENCES

1. R. Courant, K. Friedrichs and H. Lewy. *Mathematische Annalen*, 100 (1928), 32.
2. D. Hilbert. *Grundzüge einer allgemeinen Theorie der linearen Integralgleichungen* (Leipzig), 1924.
3. W. V. D. Hodge. *Proceedings of the London Mathematical Society* (2), 36 (1932), 257.
4. W. V. D. Hodge. *Proceedings of the London Mathematical Society* (2), 38 (1933), 72.
5. W. V. D. Hodge. *Proceedings of the London Mathematical Society* (2), 41 (1936), 484.
6. G. Kowalewski. *Integralgleichungen* (Berlin), 1930.
7. H. Weyl. *Die Idee der Riemannschen Fläche* (Leipzig), 1923.

Chapter IV

APPLICATIONS TO ALGEBRAIC VARIETIES

In this chapter we make use of the results obtained in previous chapters to investigate properties of algebraic varieties. We first show how to construct a set of Riemannian manifolds corresponding to an algebraic variety V_m of m complex dimensions, and then proceed to investigate the properties of the harmonic integrals on one of these manifolds. The metric on the manifold has certain special properties which enable us to make a classification of the harmonic integrals of multiplicity p into a number of sets. This classification is closely related to the classification of p-cycles of the manifold due to Lefschetz. It appears that, for each value of p not exceeding m, there is a special linear sub-set of the harmonic p-fold integrals, containing $R_p - R_{p-2}$ linearly independent integrals. We call the integrals of this sub-set the effective p-fold integrals, and show that all the important properties of the harmonic integrals on the manifold can be expressed as properties of the effective integrals.

The results we arrive at can be expressed most simply when we make use of complex parameters $(z_1, \ldots, z_m)$ on V_m, and take $(z_1, \ldots, z_m; \bar{z}_1, \ldots, \bar{z}_m)$ as parameters on the Riemannian manifold, where $\bar{z}_h$ is the conjugate imaginary of z_h. When we do this, we find that we can classify the $R_p - R_{p-2}$ effective p-fold integrals into $(p+1)$ classes. The integrals of the $(k+1)$th class can be written in the form

$$\int A_{i_1 \ldots i_{p-k} j_1 \ldots j_k} dz^{i_1} \ldots dz^{i_{p-k}} d\bar{z}^{j_1} \ldots d\bar{z}^{j_k},$$

where the summations are from 1 to m, and the integrals of the $(p-k+1)$th set are the conjugate imaginaries of the integrals of the $(k+1)$th set. The first set consists of the algebraic integrals of the first kind of multiplicity p associated with V_m, but the second, ..., pth sets depend on the particular Riemannian manifold associated with V_m which we have considered. It is shown, however, that the number of integrals in each set, and the period matrices of the integrals of each set, are relative birational invariants of the variety V_m. We then show how a number of geometrical properties of V_m can be expressed as properties of these invariant matrices.

36. Algebraic varieties. It is well known that, given any irreducible curve U_1, there is associated with it a real analytic manifold M of two dimensions, which we call its Riemann surface. This manifold is not a Riemannian manifold, since it does not carry a metric, but, as we shall see later, it is possible to attach to M a Riemannian metric so that the harmonic integrals defined by this metric on M are well-known invariants connected with the Riemann surface. We wish to generalise the idea of a Riemann surface, and to associate with any irreducible algebraic variety U_m of m dimensions a real analytic Riemannian manifold of $2m$ dimensions. We shall then study the harmonic integrals on this manifold.

If U_m lies in the projective space S_k of k dimensions, we shall show in the next paragraph how we can construct the Riemannian manifold R which is to be associated with S_k. In R there is an analytic locus M of $2m$ dimensions representing the points of S_k which lie on U_m. Now, if U_m has no multiple points, we can show that M is a Riemannian manifold, provided we define a metric on it, and this we can do by using the metric on R. It is this manifold M which we shall take to be the Riemannian manifold associated with U_m, and we shall see that, in the case $m = 1$, M is, essentially, the Riemann surface of U_1 as defined in textbooks on algebraic functions (cf. the remarks in § 1). But if U_m has multiple points, M will have singular points, and in general it does not conform to the requirements of a Riemannian manifold. There is a well-known way of getting over this difficulty when $m = 1$, but this does not help us much in considering general values of m. We must therefore consider how we should proceed in the case in which U_m has multiple points. We might try to extend the idea of a Riemannian manifold to allow the presence of singular points, and try to extend the results of previous chapters to the more extended type of manifold, but it can be readily seen that this would lead to serious complications, and indeed this method seems impracticable.

The alternative is to try to eliminate the multiple points of U_m by means of a birational transformation. It is known that an irreducible curve, or surface, can be birationally transformed into a curve, or surface, without singular points. The theorem that an algebraic variety U_m $(m > 2)$ can be transformed birationally into a variety V_m, lying in a space S_r of r dimensions, which has no multiple points, has not yet been proved satisfactorily, but there are many reasons for believing that it is true. The procedure we shall adopt in this chapter will be to assume the truth of this theorem, and replace U_m by V_m, and take M to be the Riemannian manifold of V_m constructed as we have indicated. It will be noticed that in fact what we are doing is to confine our considerations, in the cases $m > 2$, to varieties U_m which can be transformed into varieties V_m without multiple points.

When $m > 1$, there is a further complication to be taken into account. Two algebraic varieties V_m, V'_m, each without multiple points, may be in birational correspondence, and yet not be in (1-1) correspondence without exception, for certain points of one may correspond to algebraic loci of dimension greater than zero on the other. The manifolds M, M', which correspond to V_m, V'_m, respectively, are therefore not necessarily homeomorphic, and hence we can obtain from the original variety U_m different manifolds M which are not homeomorphic. The results which we obtain by means of harmonic integrals considered on M will not necessarily be the same as those obtained by considering M', and hence the invariants which we obtain are at most "relative" invariants, that is, they are common only to varieties obtainable from one another by birational transformations in which there are no exceptional elements.

We must also make it clear that the metric which we introduce on M is not uniquely fixed by the representation of U_m by a variety V_m without multiple points, even when we confine ourselves to a class of varieties V_m which are in (1-1) correspondence without exception. Thus, it would seem that by the introduction of harmonic integrals we should arrive at characters of V_m which are not even relative invariants. However, it will be shown that, from the results which we arrive at by the use of harmonic integrals, we can extract certain properties which do not depend on the metric, but are birationally invariant properties of V_m relative to the system of prime sections of V_m. By a proper system on V_m, we shall understand any continuous system of varieties of $m-1$ dimensions on V_m containing a linear system without base elements which can be used to represent V_m projectively as a variety without multiple points. We thus obtain invariants of V_m relative to each proper system. The importance of invariants relative to a proper system lies in the fact that the classification of the cycles of M, due to Lefschetz [7], is also birationally invariant relative to a proper system, and one of the main objects of this chapter is to obtain a parallelism between the results obtained by means of harmonic integrals and Lefschetz's classification of cycles.

37·1. Construction of the Riemannian manifold. Let us consider an irreducible algebraic variety V_m of m dimensions, without multiple points, lying in a projective space S_r of r dimensions. We shall construct Riemannian manifolds R corresponding to S_r, and M corresponding to V_m. A very convenient method of representing the points of S_r by the points of a real analytic locus of $2r$ dimensions has been given by Mannoury [9].

Let $(z_0, \ldots, z_r)$ be a set of complex homogeneous coordinates in S_r. Since only the ratios of the coordinates of a point are determined, we can, without loss of generality, suppose that the coordinates of each point are subject to the condition

$$z_0\bar{z}_0 + \dots + z_r\bar{z}_r = 1, \tag{1}$$

where z_h and $\bar{z}_h$ are conjugate complex numbers. Let X_h, $X_{hk} = X_{kh}$, $Y_{hk} = -Y_{kh}$ $(h \neq k)$ be rectangular cartesian coordinates in a Euclidean space E of $(r+1)^2$ dimensions, and consider the locus R defined in E by the equations

$$\left.\begin{aligned} X_h &= \sqrt{2}\, z_h\bar{z}_h, \\ X_{hk} &= z_h\bar{z}_k + \bar{z}_h z_k, \\ Y_{hk} &= i(z_h\bar{z}_k - \bar{z}_h z_k), \end{aligned}\right\} \tag{2}$$

where i is a square root of -1. R is a real locus. Moreover, since

$$\begin{aligned} \sum_{h=0}^{m} X_h^2 + \sum_{\substack{h,k=0 \\ h\neq k}}^{m} X_{hk}^2 + \sum_{h,k=0}^{m} Y_{hk}^2 &= 2\sum_h z_h^2\bar{z}_h^2 + 4\sum_{h\neq k} z_h\bar{z}_h z_k\bar{z}_k \\ &= 2\left[\sum_{h=0}^{m} z_h\bar{z}_h\right]^2 \\ &= 2, \end{aligned}$$

the locus is finite. If P is any point of S_r, and $(z_0, \dots, z_r)$ are coordinates of P satisfying (1), the most general coordinates of P which satisfy (1) are $(e^{i\theta}z_0, \dots, e^{i\theta}z_r)$, where θ is real. But, since

$$\sqrt{2}\,(e^{i\theta}z_h)\,(e^{-i\theta}\bar{z}_h) = \sqrt{2}\,z_h\bar{z}_h,$$

$$(e^{i\theta}z_h)\,(e^{-i\theta}\bar{z}_k) + (e^{-i\theta}\bar{z}_h)\,(e^{i\theta}z_k) = z_h\bar{z}_k + \bar{z}_h z_k,$$

$$i[(e^{i\theta}z_h)\,(e^{-i\theta}\bar{z}_k) - (e^{-i\theta}\bar{z}_h)\,(e^{i\theta}z_k)] = i(z_h\bar{z}_k - \bar{z}_h z_k),$$

there is a unique point on R corresponding to each point of S_r. Conversely, if $(z_0, \dots, z_r)$ and $(z_0', \dots, z_r')$ both satisfy (1), and determine the same point on R, we have

$$\frac{z_h'}{z_k'} = \frac{X_{hk} - iY_{hk}}{\sqrt{2}\,X_k} = \frac{z_h}{z_k},$$

and hence each point of R corresponds to a unique point of S_r;

R is therefore of $2r$ dimensions. In R there is a locus M which is in (1-1) correspondence with the points of S_r on V_m; M is of $2m$ dimensions. We shall prove that R and M are Riemannian manifolds. Since S_r can itself be regarded as a special case of a variety without multiple points lying in S_r, it will be sufficient if we prove the result for M.

37·2. M is a locus in E defined by a finite number of analytic equations. We first show that it has no singular points. Since V_m is an algebraic variety without singular points, we can express the neighbourhood of any point of it parametrically, as follows:

$$\rho z_h = f_h(u_1, \ldots, u_m),$$

where the Jacobian matrix of the functions $f_h(u_1, \ldots, u_m)$ is of rank m. If we write

$$u_h = x_h + ix_{m+h}$$

and substitute in (1), we can determine an analytic function ρ, different from zero, so that (1) is satisfied. Substituting in (2), we obtain an expression for M in terms of parameters $(x_1, \ldots, x_{2m})$. The Jacobian matrix is of rank $2m$, and hence M has no singular points. M is therefore a manifold of class ω, as defined in Chapter I.

To prove that M is orientable, we consider the set of all coordinate systems on it obtained from local complex parameters on V_m, as $(x_1, \ldots, x_{2m})$ is obtained from $(u_1, \ldots, u_m)$, above. Let $(u_1, \ldots, u_m)$ and $(u'_1, \ldots, u'_m)$ be two systems of complex parameters on V_m which are valid in the same domain. If

$$u_h = x_h + ix_{m+h} \qquad (h = 1, \ldots, m),$$

and

$$u'_h = x'_h + ix'_{m+h} \qquad (h = 1, \ldots, m),$$

$(x_1, \ldots, x_{2m})$ and $(x'_1, \ldots, x'_{2m})$ are local coordinate systems on M valid in the same domain. In this domain we have

$$x'_h + ix'_{m+h} = \phi_h(x_1 + ix_{m+1}, \ldots, x_m + ix_{2m}) \qquad (h = 1, \ldots, m),$$

and the Cauchy-Riemann equations are

$$\frac{\partial x'_h}{\partial x_k} = \frac{\partial x'_{m+h}}{\partial x_{m+k}}, \quad \frac{\partial x'_{m+h}}{\partial x_k} = -\frac{\partial x'_h}{\partial x_{m+k}} \qquad (h, k = 1, \ldots, m).$$

Write
$$\frac{\partial x'_h}{\partial x_k} = a_{hk}, \quad \frac{\partial x'_{m+h}}{\partial x_k} = b_{hk}.$$

Then the Jacobian matrix of the transformation of the parameters is

$$\left(\frac{\partial x'_i}{\partial x_j}\right) = \begin{pmatrix} \mathbf{a} & -\mathbf{b} \\ \mathbf{b} & \mathbf{a} \end{pmatrix},$$

and is non-singular.

Now if $\mathbf{I}_m$ is the unit matrix of m rows and columns,

$$\begin{pmatrix} \mathbf{a} & -\mathbf{b} \\ \mathbf{b} & \mathbf{a} \end{pmatrix}\begin{pmatrix} \mathbf{I}_m & i\mathbf{I}_m \\ i\mathbf{I}_m & \mathbf{I}_m \end{pmatrix} = \begin{pmatrix} \mathbf{a}-i\mathbf{b} & i(\mathbf{a}+i\mathbf{b}) \\ i(\mathbf{a}-i\mathbf{b}) & \mathbf{a}+i\mathbf{b} \end{pmatrix}$$
$$= \begin{pmatrix} \mathbf{I}_m & i\mathbf{I}_m \\ i\mathbf{I}_m & \mathbf{I}_m \end{pmatrix}\begin{pmatrix} \mathbf{a}-i\mathbf{b} & \mathbf{0} \\ \mathbf{0} & \mathbf{a}+i\mathbf{b} \end{pmatrix}.$$

Since
$$\begin{vmatrix} \mathbf{I}_m & i\mathbf{I}_m \\ i\mathbf{I}_m & \mathbf{I}_m \end{vmatrix} = 2^m \neq 0,$$

we have
$$\left|\frac{\partial x'_i}{\partial x_j}\right| = |\mathbf{a}+i\mathbf{b}|\,|\mathbf{a}-i\mathbf{b}| = N^2 > 0,$$

where N is the modulus of the Jacobian $\left|\dfrac{\partial u'_i}{\partial u_j}\right|$. The coordinate systems on M obtained from the complex parameters on V_m therefore form sets serving to orient M, as in §4. Moreover, since V_m is irreducible, M is connected.

M is therefore an analytic orientable manifold. To make it a Riemannian manifold we must introduce a metric. Since M is a locus in the Euclidean space E, given locally by the analytic equations

$$X_h = f_h(x_1, \ldots, x_{2m}),$$
$$X_{hk} = f_{hk}(x_1, \ldots, x_{2m}),$$
$$Y_{hk} = \bar{f}_{hk}(x_1, \ldots, x_{2m}),$$

we can define the metric $g_{ij}\,dx^i dx^j$ by the equations

$$g_{ij} = \sum_{h=0}^{m} \frac{\partial f_h}{\partial x_i}\frac{\partial f_h}{\partial x_j} + \sum_{\substack{h,k=0\\ h\neq k}}^{m} \frac{\partial f_{hk}}{\partial x_i}\frac{\partial f_{hk}}{\partial x_j} + \sum_{\substack{h,k=0\\ h\neq k}}^{m} \frac{\partial \bar{f}_{hk}}{\partial x_i}\frac{\partial \bar{f}_{hk}}{\partial x_j}.$$

It may be observed that, in proving the existence theorem for harmonic integrals on M, we may take the distance function $r(x, y)$ to be the Euclidean distance in E between the points of M. Then $r(x, y)$ is an analytic function, and the harmonic tensors on M have analytic components. It is also to be remembered that the metric on M depends on the representation of V_m as a variety in S_r and on the homogeneous coordinates used in S_r, so that the harmonic integrals depend on these considerations. This objection will, however, be removed later.

37·3. In the case $m = 1$, M is, essentially, the Riemann surface of V_1. Let T dênote the Riemann surface of V_1, constructed in the usual manner. If P is any place on T and $t = \sigma + i\tau$ is the local complex parameter at P on T, we may take $u_1 = t$. Then M is given by the analytic equations

$$X_h = f_h(\sigma, \tau),$$
$$X_{hk} = f_{hk}(\sigma, \tau),$$
$$Y_{hk} = g_{hk}(\sigma, \tau),$$

where the matrix

$$\begin{pmatrix} \dfrac{\partial f_h}{\partial \sigma} & \dfrac{\partial f_{hk}}{\partial \sigma} & \dfrac{\partial g_{hk}}{\partial \sigma} \\ \dfrac{\partial f_h}{\partial \tau} & \dfrac{\partial f_{hk}}{\partial \tau} & \dfrac{\partial g_{hk}}{\partial \tau} \end{pmatrix}$$

is of rank *two*. M is therefore analytically homeomorphic to T. Moreover, as we shall see below, the metric is of the form $\lambda[d\sigma^2 + d\tau^2]$. Hence the Riemann surface defined by the Riemannian manifold, as in § 1, is the Riemann surface of the curve V_1.

38·1. **Discussion of the metric.** We now examine the metrics which we have defined on R and M. We begin with R.

Let $(z_0, \ldots, z_r)$ and $(\zeta_0, \ldots, \zeta_r)$ be two points of S_r, whose coordinates satisfy (1). The distance ρ between the corresponding points of E is given by

$$\begin{aligned}\rho^2 &= 2\sum_h (z_h\bar{z}_h - \zeta_h\bar{\zeta}_h)^2 + \sum_{h\neq k}(z_h\bar{z}_k + \bar{z}_h z_k - \zeta_h\bar{\zeta}_k - \bar{\zeta}_h\zeta_k)^2 \\ &\qquad\qquad - \sum_{h\neq k}(z_h\bar{z}_k - \bar{z}_h z_k - \zeta_h\bar{\zeta}_k + \bar{\zeta}_h\zeta_k)^2 \\ &= 2\sum_h (z_h\bar{z}_h - \zeta_h\bar{\zeta}_h)^2 + 4\sum_{h\neq k}(z_h\bar{z}_k - \zeta_h\bar{\zeta}_k)(\bar{z}_h z_k - \bar{\zeta}_h\zeta_k) \\ &= 2\sum_h z_h^2\bar{z}_h^2 + 2\sum_h \zeta_h^2\bar{\zeta}_h^2 - 4\sum_h z_h\bar{z}_h\zeta_h\bar{\zeta}_h \\ &\qquad + 4\sum_{h\neq k} z_h\bar{z}_h z_k\bar{z}_k + 4\sum_{h\neq k}\zeta_h\bar{\zeta}_h\zeta_k\bar{\zeta}_k - 4\sum_h\Big[z_h\bar{\zeta}_h\sum_{k\neq h}\bar{z}_k\zeta_k\Big] \\ &= 2\Big[\sum_h z_h\bar{z}_h\Big]^2 + 2\Big[\sum_h \zeta_h\bar{\zeta}_h\Big]^2 - 4\Big[\sum_h z_h\bar{\zeta}_h\Big]\Big[\sum_h \bar{z}_h\zeta_h\Big] \\ &= 4\Big[1 - \Big|\sum_h z_h\bar{\zeta}_h\Big|^2\Big].\end{aligned}$$

Let us suppose that z_0 and ζ_0 are both different from zero, and introduce non-homogeneous coordinates

$$\left(\frac{z_1}{z_0}, \ldots, \frac{z_r}{z_0}\right), \quad \left(\frac{\zeta_1}{\zeta_0}, \ldots, \frac{\zeta_r}{\zeta_0}\right).$$

Write $$\frac{z_h}{z_0} = x_h + ix_{r+h}, \quad \frac{\zeta_h}{\zeta_0} = \xi_h + i\xi_{r+h}.$$

If we are given the non-homogeneous coordinates of the points, we can determine homogeneous coordinates which satisfy (1) by writing

$$z_0 = z_0, \quad z_h = z_0(x_h + ix_{r+h}),$$

and $$\zeta_0 = \zeta_0, \quad \zeta_h = \zeta_0(\xi_h + i\xi_{r+h}),$$

where z_0 and ζ_0 are real and positive, and satisfy

$$z_0^2\left[1 + \sum_1^{2r} x_h^2\right] = 1,$$

$$\zeta_0^2\left[1 + \sum_1^{2r} \xi_h^2\right] = 1.$$

Then

$$\rho^2 = 4\left[1 - \frac{\left|\sum_{h=1}^{r}(x_h + ix_{r+h})(\xi_h - i\xi_{r+h}) + 1\right|^2}{\left[1+\sum_1^{2r} x_h^2\right]\left[1+\sum_1^{2r}\xi_h^2\right]}\right]$$

$$= \frac{4}{\left[1+\sum_1^{2r} x_h^2\right]\left[1+\sum_1^{2r}\xi_h^2\right]}\left[\left(1+\sum_1^{2r} x_h^2\right)\left(1+\sum_1^{2r}\xi_h^2\right) - \left(1+\sum_1^{2r} x_h\xi_h\right)^2 - \left\{\sum_1^{r}(x_h\xi_{r+h} - x_{r+h}\xi_h)\right\}^2\right]$$

$$= \frac{4}{\left[1+\sum_1^{2r} x_h^2\right]\left[1+\sum_1^{2r}\xi_h^2\right]}\left[\sum_1^{2r}(x_h - \xi_h)^2 + \sum_{h,k=1}^{2r}(x_h\xi_k - x_k\xi_h)^2 - \left\{\sum_1^{r}(x_h\xi_{r+h} - x_{r+h}\xi_h)\right\}^2\right],$$

where, in the summation over h, k, each pair is counted once.

Let us now suppose that the points are near together, and write

$$x_h = \xi_h + d\xi^h.$$

To the second order of small quantities, the distance $\rho = ds$ is given by

$$ds^2 = \frac{4}{\left[1+\sum_1^{2r}\xi_h^2\right]^2}\left[\sum_1^{2r}(d\xi^h)^2 + \sum_{h,k}(\xi_h d\xi^k - \xi_k d\xi^h)^2 - \left\{\sum_1^{r}(\xi_h d\xi^{r+h} - \xi_{r+h} d\xi^h)\right\}^2\right]$$

$$= \eta_{ij}\, d\xi^i d\xi^j, \tag{3}$$

where

$$\eta_{ij} = \eta_{r+i\,r+j} = \frac{4\delta_{ij}}{\left[1+\sum_1^{2r}\xi_h^2\right]} - \frac{4[\xi_i\xi_j + \xi_{r+i}\xi_{r+j}]}{\left[1+\sum_1^{2r}\xi_h^2\right]^2} \qquad (i, j \leqslant r),$$

and

$$\eta_{i\,r+j} = -\eta_{r+i\,j} = -\frac{4[\xi_i\xi_{r+j} - \xi_{r+i}\xi_j]}{\left[1+\sum_1^{2r}\xi_h^2\right]} \qquad (i, j \leqslant r).$$

There is a convenient rule for expressing these formulae for η_{ij}. Let us suppose that ξ_i is defined for all integral values of the suffix i, in terms of $\xi_1, \ldots, \xi_{2r}$, by the formula

$$\xi_{2r+i} = -\xi_i.$$

If we define a function ψ by the equation

$$\psi = \log\left[1 + \sum_1^{2r} \xi_h^2\right],$$

we find that
$$\eta_{ij} = \frac{\partial^2 \psi}{\partial \xi_i \partial \xi_j} + \frac{\partial^2 \psi}{\partial \xi_{r+i} \partial \xi_{r+j}}$$

for all values of i, j.

38·2. We have now found the form of the metric on R in a particular set of local parameters. Let us make a transformation of the complex parameters on S_r, say,

$$\xi_h + i\xi_{r+h} = f_h(\xi'_1 + i\xi'_{r+1}, \ldots, \xi'_r + i\xi'_{2r}).$$

Defining ξ'_i, for all values of the suffix, by the equations

$$\xi'_{2r+i} = -\xi'_i,$$

we can write the Cauchy-Riemann equations in the form

$$\frac{\partial \xi_h}{\partial \xi'_k} = \frac{\partial \xi_{r+h}}{\partial \xi'_{r+k}} \qquad (h, k = 1, \ldots, 2r).$$

Then

$$\begin{aligned}
\eta_{ij} \frac{\partial \xi_i}{\partial \xi'_h} \frac{\partial \xi_j}{\partial \xi'_k} &= \frac{\partial^2 \psi}{\partial \xi_i \partial \xi_j} \frac{\partial \xi_i}{\partial \xi'_h} \frac{\partial \xi_j}{\partial \xi'_k} + \frac{\partial^2 \psi}{\partial \xi_{r+i} \partial \xi_{r+j}} \frac{\partial \xi_i}{\partial \xi'_h} \frac{\partial \xi_j}{\partial \xi'_k} \\
&= \frac{\partial^2 \psi}{\partial \xi_i \partial \xi_j} \frac{\partial \xi_i}{\partial \xi'_h} \frac{\partial \xi_j}{\partial \xi'_k} + \frac{\partial^2 \psi}{\partial \xi_{r+i} \partial \xi_{r+j}} \frac{\partial \xi_{r+i}}{\partial \xi'_{r+h}} \frac{\partial \xi_{r+j}}{\partial \xi'_{r+k}} \\
&= \frac{\partial^2 \psi}{\partial \xi'_h \partial \xi'_k} - \frac{\partial \psi}{\partial \xi_i} \frac{\partial^2 \xi_i}{\partial \xi'_h \partial \xi'_k} + \frac{\partial^2 \psi}{\partial \xi'_{r+h} \partial \xi'_{r+k}} - \frac{\partial \psi}{\partial \xi_i} \frac{\partial^2 \xi_i}{\partial \xi'_{r+h} \partial \xi'_{r+k}} \\
&= \frac{\partial^2 \psi}{\partial \xi'_h \partial \xi'_k} + \frac{\partial^2 \psi}{\partial \xi'_{r+h} \partial \xi'_{r+k}},
\end{aligned}$$

since
$$\frac{\partial^2 \xi_i}{\partial \xi'_h \partial \xi'_k} + \frac{\partial^2 \xi_i}{\partial \xi'_{r+h} \partial \xi'_{r+k}} = 0.$$

In applying the summation convention, we have of course summed for i, j, etc., from 1 to $2r$.

In the new system of coordinates the metric $\eta'_{ij} d\xi'^i d\xi'^j$ is therefore given by

$$\eta'_{ij} = \frac{\partial^2 \psi}{\partial \xi'_i \partial \xi'_j} + \frac{\partial^2 \psi}{\partial \xi'_{r+i} \partial \xi'_{r+j}}.$$

In particular, if we take new non-homogeneous coordinates to be $\left(\frac{z_0}{z_1}, \ldots, \frac{z_r}{z_1}\right)$ in S_r, the metric is given by this formula. We might, however, have begun with this system of non-homogeneous coordinates, and we should then have obtained

$$\eta'_{ij} = \frac{\partial^2 \psi'}{\partial \xi'_i \partial \xi'_j} + \frac{\partial^2 \psi'}{\partial \xi'_{r+i} \partial \xi'_{r+j}},$$

where

$$\psi' = \log\left(1 + \sum_1^{2r} \xi'^2_h\right)$$

$$= \psi - \log(\xi_1^2 + \xi_{r+1}^2).$$

This emphasises the fact that ψ is not unique. We can, indeed, modify ψ by the addition of any function ϕ which satisfies the equations

$$\frac{\partial^2 \phi}{\partial \xi_i \partial \xi_j} + \frac{\partial^2 \phi}{\partial \xi_{r+i} \partial \xi_{r+j}} = 0 \qquad (i, j = 1, \ldots, 2r).$$

In particular, we see that in the neighbourhood of any given point of R we can modify ψ so that it is analytic in the neighbourhood.

38·3. We can now consider the metric on M, which is a sub-manifold of R. Let V_m be given locally by the parametric equations

$$\xi_h + i\xi_{r+h} = f_h(x_1 + i x_{m+1}, \ldots, x_m + i x_{2m}) \quad (h = 1, \ldots, r),$$

and now define x_h for all integral values of the suffix h by the equations

$$x_{2m+h} = -x_h.$$

Just as when we considered the transformation of local coordinates in S_r, so we now prove that the metric on M is given by

$$g_{ij}\,dx^i\,dx^j,$$

where
$$g_{ij} = \frac{\partial^2\psi}{\partial x_i \partial x_j} + \frac{\partial^2\psi}{\partial x_{m+i}\partial x_{m+j}} \qquad (i, j = 1, \ldots, 2m). \qquad (4)$$

The metric on M will have this form whenever the local coordinates on M are deduced from the complex parameters on V_m in the way we have done above. In this chapter we shall not have to consider parametric systems on V_m obtained in any other way, and we may restrict the systems of allowable coordinates to systems like $(x_1, \ldots, x_{2m})$, where $(x_1 + ix_{m+1}, \ldots, x_m + ix_{2m})$ are complex parameters on V_m. For these restricted systems of allowable coordinates we shall prove certain theorems which are not true for general coordinate systems obtained by analytic transformations. Since we shall not usually state explicitly that the theorems only hold for the restricted systems of coordinates, we must warn the reader not to apply the results of this chapter to more general coordinate systems on M. In the present chapter it is to be understood that the term "allowable coordinate systems" is to be interpreted in the restricted sense. We observe that when $m = 1$, the metric in our allowable coordinate system is always of the form $\lambda\,[(dx^1)^2 + (dx^2)^2]$, and hence the Riemann surface determined from the manifold, as in § 1, is the Riemann surface of V_1.

We shall refer to the Riemannian manifold M which we have constructed as an *algebraic manifold*.

39·1. The affine connection and curvature tensor. A Riemannian space M in which the metric is given by (4) has some special properties which are of interest; they will play an important part in our later investigations.

We first show that at any point of M we can find an allowable coordinate system of the restricted type which is geodesic.

Let $(z_0, \ldots, z_r)$ and $(\tilde{z}_0, \ldots, \tilde{z}_r)$ be two systems of homogeneous coordinates in S_r. Using either of these systems, we can follow the process of § 37 to construct a Riemannian manifold of S_r (and hence of V_m), and the two Riemannian manifolds will be congruent if and only if

$$|z'_0\bar{z}_0 + \ldots + z'_r\bar{z}_r| = |\tilde{z}'_0\bar{\tilde{z}}_0 + \ldots + \tilde{z}'_r\bar{\tilde{z}}_r|,$$

where $(z_0, \ldots, z_r)$, $(z'_0, \ldots, z'_r)$ are the coordinates (satisfying (1)) of two points of S_r referred to the first coordinate system, and $(\tilde{z}_0, \ldots, \tilde{z}_r)$, $(\tilde{z}'_0, \ldots, \tilde{z}'_r)$ are the coordinates (also satisfying (1)) of the same two points referred to the second coordinate system. Now, if the transformation between the coordinate systems is

$$\tilde{z}_h = \sum_{k=0} a_{hk} z_k \qquad (h = 0, \ldots, r),$$

the necessary and sufficient condition that the manifolds are congruent is

$$|z'_0\bar{z}_0 + \ldots + z'_r\bar{z}_r| = |\sum_{h,k} a_{0h}\bar{a}_{0k} z'_h \bar{z}_k + \ldots + \sum_{h,k} a_{rh}\bar{a}_{rk} z'_h \bar{z}_k|.$$

Hence we obtain the conditions

$$\sum_i a_{ih}\bar{a}_{ik} = \delta^h_k e^{i\alpha};$$

in particular, if $\mathbf{a} = (a_{hk})$ is a unitary matrix, that is, if

$$\mathbf{a}'\bar{\mathbf{a}} = \mathbf{I},$$

the condition is satisfied. When the Riemannian manifolds are congruent, they are clearly equivalent for all our purposes. We now construct a manifold $\tilde{M}$ equivalent to M on which we can find a coordinate system which is geodesic at the point corresponding to a given point P of V_m.

Let P be the point $(\alpha_0, \ldots, \alpha_r)$ in the coordinate system $(z_0, \ldots, z_r)$, where

$$\alpha_0\bar{\alpha}_0 + \ldots + \alpha_r\bar{\alpha}_r = 1,$$

and let the tangent space to V_m at P be given by the equations

$$\Sigma \beta_{ij} z_j = 0 \qquad (i=1, \ldots, r-m).$$

Let
$$\gamma_{1i} = \lambda_1 \beta_{1i},$$

where
$$\lambda_1 \bar{\lambda}_1 \sum_i \beta_{1i} \bar{\beta}_{1i} = 1,$$

that is,
$$\sum_i \gamma_{1i} \bar{\gamma}_{1i} = 1.$$

Next let
$$\gamma_{2i} = \lambda_2(\beta_{2i} + \mu_{21} \gamma_{1i}),$$

where
$$\sum_i \beta_{2i} \bar{\gamma}_{1i} + \mu_{21} = 0,$$

and
$$\lambda_2 \bar{\lambda}_2 \sum_i (\beta_{2i} + \mu_{21} \gamma_{1i})(\bar{\beta}_{2i} + \bar{\mu}_{2i} \bar{\gamma}_{1i}) = 1,$$

that is
$$\sum_i \gamma_{2i} \bar{\gamma}_{2i} = 1, \quad \sum_i \gamma_{2i} \bar{\gamma}_{1i} = 0.$$

Proceeding in this way, we can replace the equations of the tangent space to V_m at P by the equations

$$\sum_j \gamma_{ij} z_j = 0 \qquad (i=1, \ldots, r-m),$$

where
$$\sum_j \gamma_{hj} \bar{\gamma}_{kj} = \delta_k^h, \qquad (h, k=1, \ldots, r-m),$$

$$\sum_j \gamma_{hj} \alpha_j = 0 \qquad (h=1, \ldots, r-m).$$

Now consider the matrix L given by

$$L = \begin{pmatrix} \alpha_0 & \bar{\gamma}_{10} & \cdots & \bar{\gamma}_{r-m\,0} \\ \vdots & & & \\ \alpha_r & \bar{\gamma}_{1r} & \cdots & \bar{\gamma}_{m-r\,r} \end{pmatrix}.$$

We can add to the right of this matrix m further columns so that the resulting matrix $\mathbf{a}$ is unitary. Then consider the new system of coordinates $(\tilde{z}_0, \ldots, \tilde{z}_r)$ given by

$$z_h = \sum_k a_{hk} \tilde{z}_k.$$

In this coordinate system, P is the point $(1, 0, \ldots, 0)$ and the tangent space is

$$\tilde{z}_h = 0 \qquad (h=1, \ldots, r-m).$$

Using the coordinate system $(\tilde{z}_0, \ldots, \tilde{z}_r)$, we construct the Riemannian manifold $\tilde{M}$ of V_m, where $\tilde{M}$ is equivalent to M. The function ψ can be taken as

$$\psi = \log\left[1 + \frac{\tilde{z}_1 \bar{\tilde{z}}_1}{\tilde{z}_0 \bar{\tilde{z}}_0} + \ldots + \frac{\tilde{z}_r \bar{\tilde{z}}_r}{\tilde{z}_0 \bar{\tilde{z}}_0}\right].$$

In the neighbourhood of P, V_m is given by

$$\frac{\tilde{z}_h}{\tilde{z}_0} = f_h\left(\frac{\tilde{z}_{r-m+1}}{\tilde{z}_0}, \ldots, \frac{\tilde{z}_r}{\tilde{z}_0}\right) \qquad (h = 1, \ldots, r-m),$$

where the functions on the right are power series beginning with terms of the second degree. Take

$$u_j = \frac{\tilde{z}_{r-m+j}}{2\tilde{z}_0} \qquad (j = 1, \ldots, m),$$

as the local parameters on V_m. Then

$$\psi = \log\left[1 + \tfrac{1}{4}\sum_i^{2m} x_i^2 + \epsilon\right],$$

where ϵ is a power series beginning with terms of the fourth degree. If we now calculate g_{ij} by (4), and then calculate Γ^i_{jk}, we find that at P

$$g_{ij} = \delta^i_j, \quad \Gamma^i_{jk} = 0.$$

The coordinates are therefore geodesic on $\tilde{M}$. But, since $\tilde{M}$ is equivalent to M, it follows that there exist on M restricted allowable coordinates which are geodesic at a given point.

39·2. In the formulae which we now give for the differential invariants of M we find it convenient to allow the indices in components to take all integral values, with the condition that the addition or subtraction of $2m$ to or from one index of a component changes the sign of this component. Thus, for example,

$$g_{i\,2m+j} = -g_{ij},$$
$$\Gamma^{2m+i}_{jk} = -\Gamma^i_{jk}.$$

It is to be understood, of course, that in applying the summation convention, the summation takes place from 1 to $2m$.

Let $T^{j_1\ldots j_q}_{i_1\ldots i_p}$ be any tensor of weight W, and let

$$S^{j_1\ldots j_q}_{i_1\ldots i_p} = T^{j_1\ldots j_{r-1}\, m+j_r\, j_{r+1}\ldots j_q}_{i_1\ldots\ldots\ldots\ldots\ldots\ldots\ldots\ldots i_p}.$$

Consider an allowable transformation of coordinates to a new system $(\bar{x}_1, \ldots, \bar{x}_{2m})$. Then

$$\left|\frac{\partial x_i}{\partial \bar{x}_j}\right|^W S^{b_1\ldots b_q}_{a_1\ldots a_p} \frac{\partial x_{a_1}}{\partial \bar{x}_{i_1}} \cdots \frac{\partial x_{a_p}}{\partial \bar{x}_{i_p}} \frac{\partial \bar{x}_{j_1}}{\partial x_{b_1}} \cdots \frac{\partial \bar{x}_{j_q}}{\partial x_{b_q}}$$

$$= \left|\frac{\partial x_i}{\partial \bar{x}_j}\right|^W T^{b_1\ldots b_{r-1}\, m+b_r\, b_{r+1}\ldots b_q}_{a_1\ldots\ldots\ldots\ldots\ldots\ldots\ldots\ldots a_p} \frac{\partial x_{a_1}}{\partial \bar{x}_{i_1}} \cdots \frac{\partial x_{a_p}}{\partial \bar{x}_{i_p}}$$

$$\times \frac{\partial \bar{x}_{j_1}}{\partial x_{b_1}} \cdots \frac{\partial \bar{x}_{j_{r-1}}}{\partial x_{b_{r-1}}} \frac{\partial \bar{x}_{m+j_r}}{\partial x_{m+b_r}} \frac{\partial \bar{x}_{j_{r+1}}}{\partial x_{b_{r+1}}} \cdots \frac{\partial \bar{x}_{j_q}}{\partial x_{b_q}}$$

$$= \bar{T}^{j_1\ldots j_{r-1}\, m+j_r\, j_{r+1}\ldots j_q}_{i_1\ldots\ldots\ldots\ldots\ldots\ldots\ldots\ldots i_p}.$$

Hence from $T^{j_1\ldots j_q}_{i_1\ldots i_p}$ we can define a new tensor by increasing one of the contravariant indices by m. There will be no confusion if we denote the tensor $S^{j_1\ldots j_q}_{i_1\ldots i_p}$ by $T^{j_1\ldots j_{r-1}\, m+j_r\, j_{r+1}\ldots j_q}_{i_1\ldots\ldots\ldots\ldots\ldots\ldots\ldots\ldots i_p}$. Thus, in the case $m = 1$, if T^{ij} is a tensor whose components in an allowable system of coordinates are

$$T^{11} = a, \quad T^{12} = b, \quad T^{21} = c, \quad T^{22} = d,$$

$T^{m+i\,j}$ has components in this system of coordinates given by

$$T^{m+1\,1} = c, \quad T^{m+1\,2} = d, \quad T^{m+2\,1} = -a, \quad T^{m+2\,2} = -b.$$

A similar result can be proved for the covariant indices, and, clearly, further tensors can be obtained by applying the process to more than one index. We notice that, on account of (4), $g_{i\,m+j}$ and $g^{i\,m+j}$ are skew-symmetric tensors.

39·3. Now let us consider the Christoffel symbols. We have

$$\Gamma^k_{ij} = \tfrac{1}{2} g^{ka}[\psi_{ija} + \psi_{i\,m+j\,m+a} + \psi_{m+i\,j\,m+a} - \psi_{m+i\,m+j\,a}],$$

where we have written

$$\psi_{ijk} = \frac{\partial^3 \psi}{\partial x_i \partial x_j \partial x_k}.$$

Then,

$$\begin{aligned}\Gamma^k_{m+i\,j} &= \tfrac{1}{2}g^{ka}[\psi_{m+i\,j\,a}+\psi_{m+i\,m+j\,m+a}-\psi_{ij\,m+a}+\psi_{i\,m+j\,a}]\\ &= \Gamma^k_{i\,m+j}\\ &= \tfrac{1}{2}g^{k\,m+a}[\psi_{m+i\,j\,m+a}-\psi_{m+i\,m+j\,a}+\psi_{ija}+\psi_{i\,m+j\,m+a}]\\ &= -\tfrac{1}{2}g^{m+k\,a}[\psi_{ija}+\psi_{i\,m+j\,m+a}+\psi_{m+i\,j\,m+a}-\psi_{m+i\,m+j\,a}]\\ &= -\Gamma^{m+k}_{ij}\end{aligned}$$

These results enable us to express the covariant derivative of a tensor $S^{j_1\ldots j_q}_{i_1\ldots i_p} = T^{j_1\ldots j_{r-1}\,m+j_r\,j_{r+1}\ldots j_q}_{i_1\ldots\ldots\ldots\ldots\ldots\ldots\ldots\ldots i_p}$ in terms of the covariant derivative of $T^{j_1\ldots j_q}_{i_1\ldots i_p}$. For convenience we shall consider the case $r = 1$. If $T^{m+j_1\,j_2\ldots j_q}_{i_1\ldots\ldots\ldots i_p,k}$ is the tensor obtained from $T^{j_1\ldots\ldots j_q}_{i_1\ldots i_p,k}$ by increasing the first contravariant suffix by m, we have

$$\begin{aligned}S^{j_1\ldots j_q}_{i_1\ldots i_p,k}-T^{m+j_1\,j_2\ldots j_q}_{i_1\ldots\ldots\ldots i_p,k} &= S^{\alpha\,j_2\ldots j_q}_{i_1\ldots\ldots i_p}\Gamma^{j_1}_{\alpha k}-T^{\alpha\,j_2\ldots j_q}_{i_1\ldots\ldots i_p}\Gamma^{m+j_1}_{\alpha k}\\ &= T^{m+\alpha\,j_2\ldots j_q}_{i_1\ldots\ldots\ldots i_p}\Gamma^{j_1}_{\alpha k}+T^{\alpha\,j_2\ldots j_q}_{i_1\ldots\ldots i_p}\Gamma^{j_1}_{m+\alpha\,k}\\ &= -T^{\alpha\,j_2\ldots j_q}_{i_1\ldots\ldots i_p}\Gamma^{j_1}_{m+\alpha\,k}+T^{\alpha\,j_2\ldots j_q}_{i_1\ldots\ldots i_p}\Gamma^{j_1}_{m+\alpha\,k}\\ &= 0.\end{aligned}$$

A similar proof holds where a covariant index of $T^{j_1\ldots j_q}_{i_1\ldots i_p}$ is increased by m. By a repetition of the argument, we see that if $(T^{j_1\ldots j_q}_{i_1\ldots i_p})^*$ is a tensor obtained by increasing any set of the indices of $T^{j_1\ldots j_q}_{i_1\ldots i_p}$ by m,

$$(T^{j_1\ldots j_q}_{i_1\ldots i_p})^*{}_{,k_1,k_2,\ldots} = (T^{j_1\ldots j_q}_{i_1\ldots i_p,k_1,k_2,\ldots})^*,$$

or, in other words, the operation of increasing an index by m is commutative with the operation of covariant differentiation.

39·4. Finally, certain properties of the tensors obtained by increasing some of the indices of the curvature tensor by m must be proved now, since they will be required later.

We may use allowable coordinates which are geodesic. Then for the Riemann-Christoffel tensor we have

$$R_{ijkl} = \tfrac{1}{2}[\psi_{m+i\,j\,m+k\,l}+\psi_{i\,m+j\,k\,m+l}-\psi_{m+i\,j\,k\,m+l}-\psi_{i\,m+j\,m+k\,l}].$$

Therefore we find immediately that

$$R_{ij\,m+k\,l} = -R_{ijk\,m+l},$$

and

$$R_{m+i\,jkl} = -R_{i\,m+j\,kl}.$$

But we have seen that $R_{ij\,m+kl}$, etc. are tensors. Hence these equations will hold in all allowable coordinate systems. Similarly, we have

$$B^a_{j\,m+kl} = g^{ai}R_{ij\,m+kl} = -g^{ai}R_{ijk\,m+l} = -B^a_{jk\,m+l},$$

and

$$\begin{aligned} B^{m+a}_{jkl} = g^{m+a\,i}R_{ijkl} &= -g^{a\,m+i}R_{ijkl} \\ &= g^{ai}R_{m+i\,jkl} \\ &= -g^{ai}R_{i\,m+j\,kl} \\ &= -B^a_{m+j\,kl}. \end{aligned}$$

40. **Harmonic integrals on an algebraic manifold.** We have now found all the properties of the differential geometry of a manifold M with a metric given by (4) § 38·3 which we shall require. Our purpose is to use these properties in order to derive properties of harmonic integrals associated with this metric, with the object of obtaining invariants of an algebraic variety of geometrical interest. We shall show that the harmonic integrals of multiplicity p on an algebraic manifold can be analysed into a number of sets in a significant way, and as a first step towards this we prove two preliminary theorems on harmonic tensors on an algebraic manifold.

THEOREM I. *If $P_{i_1 \ldots i_p}$ is a harmonic tensor, then*

$$Q_{i_1 \ldots i_{p-2}} = g^{r\,m+s}P_{rs\,i_1 \ldots i_{p-2}}$$

is also a harmonic tensor, or else it is zero.

To prove this, let

$$L_{i_1 \ldots i_{p-1}} = \sum_{r=1}^{p-1}(-1)^{r-1}Q_{i_1 \ldots i_{r-1}\,i_{r+1} \ldots i_{p-1},\,i_r}.$$

Since

$$\sum_{1}^{p+1}(-1)^{r-1}P_{i_1 \ldots i_{r-1}\,i_{r+1} \ldots i_{p+1},\,i_r} = 0,$$

we have

$$L_{i_1 \ldots i_{p-1}} = 2g^{r\,m+s}P_{r\,i_1 \ldots i_{p-1},\,s}.$$

Now $$\sum_1^p (-1)^{r-1} L_{i_1\ldots i_{r-1} i_{r+1}\ldots i_p, i_r} = 0.$$

Also,

$$\begin{aligned} g^{ij} L_{i_1\ldots i_{p-2} i, j} &= 2g^{ij} g^{r\, m+s} P_{r\, i_1\ldots i_{p-2} i, s, j} \\ &= -2g^{ij} g^{r\, m+s} [P_{\alpha\, i_1\ldots i_{p-2} i} B^\alpha_{rsj} + P_{r\, i_1\ldots i_{p-2} \alpha} B^\alpha_{isj}] \\ &\quad - 2g^{ij} g^{r\, m+s} \sum_{t=1}^{p-2} P_{r\, i_1\ldots i_{t-1} \alpha\, i_{t+1}\ldots i_{p-2} i} B^\alpha_{i_t sj}. \end{aligned}$$

This is proved by changing the order of covariant differentiation and using the fact that $P_{i_1\ldots i_p}$ is harmonic (§ 35). Now

$$\begin{aligned} g^{ij} g^{r\, m+s} P_{\alpha\, i_1\ldots i_{p-2} i} B^\alpha_{rsj} &= g^{rs} g^{i\, m+j} P_{\alpha\, i_1\ldots i_{p-2} r} B^\alpha_{ijs} \\ &= -g^{rs} g^{ij} P_{\alpha\, i_1\ldots i_{p-2} r} B^\alpha_{i\, m+j\, s} \\ &= -g^{rs} g^{ij} P_{\alpha\, i_1\ldots i_{p-2} r} B^\alpha_{i\, m+s\, j} \\ &= -g^{r\, m+s} g^{ij} P_{r\, i_1\ldots i_{p-2} \alpha} B^\alpha_{isj}. \end{aligned}$$

Also $$g^{rj} g^{is} B^\alpha_{\beta\, m+s\, j} = g^{rj} g^{is} B^\alpha_{\beta\, m+j\, s} = g^{rs} g^{ij} B^\alpha_{\beta\, m+s\, j}.$$

Hence $g^{ij} g^{r\, m+s} B^\alpha_{\beta sj}$ is symmetrical in r and i. Since

$$P_{r\, i_1\ldots i_{t-1} \alpha\, i_{t+1}\ldots i_{p-2} i}$$

is skew-symmetric in r and i, we conclude that

$$g^{ij} L_{i_1\ldots i_{p-2} i, j} = 0,$$

and hence $$L = \frac{1}{(p-1)!} L_{i_1\ldots i_{p-1}} dx^{i_1} \ldots dx^{i_{p-1}}$$

is harmonic. But L is a null form; and hence, by § 29, it must be zero. Hence

$$Q = \frac{1}{(p-2)!} Q_{i_1\ldots i_{p-2}} dx^{i_1} \ldots dx^{i_{p-2}} \to 0.$$

Also, $$g^{ij} Q_{i_1\ldots i_{p-3} i, j} = g^{ij} g^{r\, m+s} P_{rs\, i_1\ldots i_{p-3} i, j} = 0,$$

since $P_{i_1\ldots i_p}$ is harmonic. The theorem is therefore proved.

Corollary. Incidentally we have proved that if $P_{i_1\ldots i_p}$ is harmonic,

$$g^{r\, m+s} P_{i_1\ldots i_{p-1} r, s} = 0.$$

THEOREM II. *If $P_{i_1\ldots i_p}$ is harmonic, so is*

$$Q_{i_1\ldots i_p} = P_{m+i_1\ldots m+i_p}.$$

Let
$$\sum_1^{p+1}(-1)^{r-1}Q_{i_1\ldots i_{r-1}i_{r+1}\ldots i_{p+1},i_r} = L_{i_1\ldots i_{p+1}}.$$

Then
$$\sum_1^{p+2}(-1)^{r-1}L_{i_1\ldots i_{r-1}i_{r+1}\ldots i_{p+2},i_r} = 0,$$

and

$$g^{ij}L_{i_1\ldots i_p i,j} = (-1)^p g^{ij}Q_{i_1\ldots i_p,i,j} + g^{ij}\sum_1^p(-1)^{r-1}Q_{i_1\ldots i_{r-1}i_{r+1}\ldots i_p i,i_r,j}$$

$$= (-1)^p\sum_1^p g^{ij}P_{m+i_1\ldots m+i_{r-1}\,i\,m+i_{r+1}\ldots m+i_p,m+i_r,j}$$

$$+(-1)^p\sum_1^p g^{ij}P_{m+i_1\ldots m+i_{r-1}\,i\,m+i_{r+1}\ldots m+i_p,i_r,m+j}$$

$$= (-1)^p\sum_1^p g^{ij}P_{m+i_1\ldots m+i_{r-1}\,i\,m+i_{r+1}\ldots m+i_p,j,m+i_r}$$

$$+(-1)^{p-1}\sum_1^p g^{ij}P_{m+i_1\ldots m+i_{r-1}\,\alpha\,m+i_{r+1}\ldots m+i_p}B^{\alpha}_{i\;m+i_r j}$$

$$+(-1)^{p-1}\sum_{\substack{r,s=1\\ r\neq s}}^p g^{ij}P_{m+i_1\ldots m+i_{r-1}\,i\,m+i_{r+1}\ldots m+i_{s-1}\,\beta\,m+i_{s+1}\ldots m+i_p} \times B^{\beta}_{m+i_s m+i_r j}$$

$$+(-1)^p\sum_1^p g^{ij}P_{m+i_1\ldots m+i_{r-1}\,i\,m+i_{r+1}\ldots m+i_p,m+j,i_r}$$

$$+(-1)^{p-1}\sum_1^p g^{ij}P_{m+i_1\ldots m+i_{r-1}\,\alpha\,m+i_{r+1}\ldots m+i_p}B^{\alpha}_{i\,i_r\,m+j}$$

$$+(-1)^{p-1}\sum_{\substack{r,s=1\\ r\neq s}}^p g^{ij}P_{m+i_1\ldots m+i_{r-1}\,i\,m+i_{r+1}\ldots m+i_{s-1}\,\beta\,m+i_{s+1}\ldots m+i_p} \times B^{\beta}_{m+i_s i_r m+j}.$$

The first term vanishes since $P_{i_1\ldots i_p}$ is harmonic, and the fourth vanishes on account of the corollary to Theorem I. The remaining terms cancel, since

$$B^{\alpha}_{i\;m+r\,j} = -B^{\alpha}_{i\,r\,m+j}.$$

Therefore the form

$$L = \frac{1}{(p+1)!}L_{i_1\ldots i_{p+1}}dx^{i_1}\ldots dx^{i_{p+1}}$$

is harmonic. But L is a null form, and therefore, by § 29, it must be zero. Hence

$$\sum_{1}^{p+1}(-1)^{r-1}Q_{i_1\ldots i_{r-1}i_{r+1}\ldots i_{p+1},i_r}=0.$$

But
$$\begin{aligned}g^{rs}Q_{i_1\ldots i_{p-1}r,s}&=g^{rs}P_{m+i_1\ldots m+i_{p-1}m+r,s}\\&=-g^{m+r\,s}P_{m+i_1\ldots m+i_{p-1}r,s}\\&=g^{r\,m+s}P_{m+i_1\ldots m+i_{p-1}\,r,s}\\&=0,\end{aligned}$$

on account of the corollary to Theorem I. Hence $Q_{i_1\ldots i_p}$ is a harmonic tensor.

41·1. The fundamental forms. The Riemannian manifold R has the Betti numbers†

$$R_0=R_2=\ldots=R_{2r}=1,$$
$$R_1=R_3=\ldots=R_{2r-1}=0.$$

If S_t is a linear t-space in S_r, there corresponds to it on R a $2t$-cycle which we denote by Γ_{2t}. Now, Γ_{2t} is the Riemannian manifold of S_t, and we orient it in the usual way, by means of complex parameters on S_t. The following two properties hold:

(i) if Γ'_{2t} is any $2t$-cycle on R,

$$\Gamma'_{2t}\sim\lambda\Gamma_{2t},$$

where λ is an integer;

(ii) $$(\Gamma_{2t}\cdot\Gamma_{2(r-t)})=(-1)^{t(r-t)}.$$

The metric on R is given, in the notation of § 38·1, equation (3), by $\eta_{ij}d\xi^i d\xi^j$.

There exists one harmonic p-fold integral for each even value of p $(0\leqslant p\leqslant 2r)$. We now wish to determine these. We must, of course, remember that we are on R, and hence, in applying the results of § 39, we must replace m by r, and determine the

† The statements concerning the topology of algebraic manifolds are quoted from the results obtained by Lefschetz [7], due attention being paid to the differences in the convention of orientation.

affine connection from (3) instead of (4). Now η_{r+ij} is a skew-symmetric absolute tensor and

$$\eta_{r+ij,k} = 0.$$

Hence
$$\eta_{r+jk,i} + \eta_{r+ki,j} + \eta_{r+ij,k} = 0,$$

and
$$\eta^{jk}\eta_{r+ij,k} = 0.$$

Therefore,
$$\nu_1 = \frac{1}{2!}\eta_{r+ij}d\xi^i d\xi^j$$

is a harmonic 2-form. In the same way, we show that

$$\nu_2 = \tfrac{1}{2}\nu_1 \times \nu_1, \quad \nu_3 = \tfrac{1}{3}\nu_1 \times \nu_2, \quad \ldots, \quad \nu_r = \frac{1}{r}\nu_1 \times \nu_{r-1}$$

are harmonic forms of multiplicity 4, 6, ..., $2r$ respectively.

The period of the integral of ν_1 on Γ_2 can be calculated by defining S_1 to be

$$\xi_2 = \ldots = \xi_r = 0, \quad \xi_{r+2} = \ldots = \xi_{2r} = 0.$$

Then
$$\int_{\Gamma_2} \nu_1 = 4\int_{S_1} \frac{d\xi^1 d\xi^{r+1}}{[1+\xi_1^2+\xi_{r+1}^2]^2} = 4\pi.$$

In the same way we can show that

$$\int_{\Gamma_{2t}} \nu_t = \frac{(4\pi)^t}{t!}(-1)^{\frac{1}{2}t(t-1)}.$$

It is easily verified that

$$\nu_r = (-1)^{\frac{1}{2}r(r-1)}\,|\,\eta_{ij}\,|^{\frac{1}{2}}\,d\xi^1 \ldots d\xi^{2r}.$$

The forms $\nu_1, \ldots, \nu_r$ define forms on the manifold M contained in R. Let us consider ν_1. This defines the form

$$\omega_1 = \frac{1}{2!}p_{ij}\,dx^i dx^j,$$

where
$$p_{ij} = \eta_{r+hk}\frac{\partial \xi_h}{\partial x_i}\frac{\partial \xi_k}{\partial x_j}$$

(where, applying the summation convention, we sum from 1 to $2r$). But, since

$$\frac{\partial \xi_h}{\partial x_i} = \frac{\partial \xi_{r+h}}{\partial x_{m+i}},$$

we obtain

$$p_{ij} = g_{m+i\,j},$$

that is,

$$\omega_1 = \frac{1}{2!} g_{m+i\,j}\, dx^i\, dx^j.$$

Similarly,

$$\omega_2 = \tfrac{1}{2}\omega_1 \times w_1, \ldots, \omega_m = \frac{1}{m}\omega_1 \times \omega_{m-1}$$

are forms on M, and as we proved that $\nu_1, \ldots, \nu_r$ are harmonic on R, so we can now prove that $\omega_1, \ldots, \omega_r$ are harmonic on M. We shall call them the *fundamental forms* on M.

41·2. Now let $\overline{\Gamma}_{2t}$ be any cycle on M; it is also a cycle on R, and on R we have

$$\overline{\Gamma}_{2t} \sim \lambda \Gamma_{2t},$$

where

$$\lambda = (-1)^{t(r-t)} (\overline{\Gamma}_{2t} \cdot \Gamma_{2(r-t)}).$$

But

$$\int_{\overline{\Gamma}_{2t}} \omega_t = \int_{\overline{\Gamma}_{2t}} \nu_t = \lambda \int_{\Gamma_{2t}} \nu_t = (-1)^{rt+\frac{1}{2}t(t+1)} (\overline{\Gamma}_{2t} \cdot \Gamma_{2(r-t)}) \frac{(4\pi)^t}{t!}.$$

In particular, if $\overline{\Gamma}_{2t}$ is a cycle corresponding to a variety V_t on V_m, of order k,

$$\lambda = k,$$

so that

$$\int_{\overline{\Gamma}_{2t}} \omega_t = k \frac{(4\pi)^t}{t!} (-1)^{\frac{1}{2}t(t-1)}.$$

If we use coordinates which are geodesic at P and evaluate at P, we have

$$\omega_1 = \sum_i dx^i\, dx^{m+i},$$

$$\omega_q = (-1)^{\frac{1}{2}q(q-1)} \sum dx^{i_1} \ldots dx^{i_q}\, dx^{m+i_1} \ldots dx^{m+i_q}.$$

Hence, if, as usual, we denote the dual of ω_q by ω_q^*,

$$\omega_q^* = (-1)^{\frac{1}{2}m(m-1)} \omega_{m-q}.$$

It may be observed that, in any allowable coordinate system,

$$\omega_q = \frac{1}{(2q)!} N^{(q)}_{i_1 \dots i_{2q}} dx^{i_1} \dots dx^{i_{2q}},$$

where $N^{(q)}_{i_1 \dots i_{2q}}$ is the Pfaffian of the skew-symmetric determinant

$$\begin{vmatrix} g_{m+i_1 i_1} & \cdots & g_{m+i_1 i_q} \\ \vdots & & \\ g_{m+i_q i_1} & \cdots & g_{m+i_q i_q} \end{vmatrix}.$$

42·1. An analysis of forms associated with an algebraic manifold. The fundamental forms which we have defined on M, together with the two theorems proved in § 40, lead to a classification of the harmonic integrals associated with M which is the basis of the application of our theory to algebraic geometry. In this paragraph we shall prove certain preliminary results concerning p-forms on an algebraic manifold, in which we do not use the condition of integrability; and hence the analysis is not confined to harmonic forms. We shall find the following notation convenient. Given the p-form

$$P = \frac{1}{p!} P_{i_1 \dots i_p} dx^{i_1} \dots dx^{i_p},$$

we define a $(p-2)$-form $P^{(1)}$ associated with it by the equation

$$P^{(1)} = \frac{1}{(p-2)!} g^{r\, m+s} P_{rs\, i_1 \dots i_{p-2}} dx^{i_1} \dots dx^{i_{p-2}} \qquad (p \geqslant 2),$$

or by

$$P^{(1)} = 0 \qquad (p < 2).$$

We then define $P^{(r)}$ by the recurrence relations

$$P^{(0)} = P, \quad P^{(r)} = [P^{(r-1)}]^{(1)}.$$

We also define an associated p-form

$$P' = \frac{1}{p!} P_{m+i_1 \dots m+i_p} dx^{i_1} \dots dx^{i_p}.$$

By § 39·2, this form is uniquely determined by P, and does not depend on the allowable coordinate system used. We note that

$$[P']' = (-1)^p P.$$

42·2. We first prove three lemmas, as aids to calculation.

Lemma A. If P is any p-form, then

$$[P \times \omega_1]^{(r)} = P^{(r)} \times \omega_1 - 2r(m-p+r-1)\,P^{(r-1)}.$$

The formula is proved by direct calculation when $r = 1$. Let us assume it true for index $r-1$. Then,

$$\begin{aligned}[P \times \omega_1]^{(r)} &= [(P \times \omega_1)^{(r-1)}]^{(1)} \\ &= [P^{(r-1)} \times \omega_1]^{(1)} - 2(r-1)(m-p+r-2)\,P^{(r-1)} \\ &= P^{(r)} \times \omega_1 - 2(m-p+2r-2)\,P^{(r-1)} \\ &\qquad\qquad - 2(r-1)(m-p+r-2)\,P^{(r-1)} \\ &= P^{(r)} \times \omega_1 - 2r(m-p+r-1)\,P^{(r-1)}.\end{aligned}$$

Lemma B. This proves a similar result:

$$[P \times \omega_q]^{(1)} = P^{(1)} \times \omega_q - 2(m-p-q+1)\,P \times \omega_{q-1}.$$

This formula holds for $q = 1$, by Lemma A. We therefore assume it true for index $q-1$, and prove it true generally by induction. By Lemma A, and the hypothesis of induction, we have

$$\begin{aligned}[P \times \omega_q]^{(1)} &= q^{-1}[P \times \omega_{q-1} \times \omega_1]^{(1)} \\ &= q^{-1}[P \times \omega_{q-1}]^{(1)} \times \omega_1 - 2q^{-1}(m-p-2q+2)\,P \times \omega_{q-1} \\ &= P^{(1)} \times \omega_q - 2q^{-1}(m-p-q+2)(q-1)\,P \times \omega_{q-1} \\ &\qquad\qquad - 2q^{-1}(m-p-2q+2)\,P \times \omega_{q-1} \\ &= P^{(1)} \times \omega_q - 2(m-p-q+1)\,P \times \omega_{q-1}.\end{aligned}$$

Lemma C. If P is any p-form ($p \leqslant m$) *such that*

$$P^{(1)} = 0,$$

then, if $p+q \leqslant m$,

$$[P \times \omega_q]^{(q)} = (-2)^q \frac{(m-p)!}{(m-p-q)!}\,P,$$

and if $p+q > m$, $\qquad [P \times \omega_q]^{(q)} = 0.$

When $q = 1$, the theorem is true by Lemma A. We may therefore proceed to prove the general result by induction. By Lemma A,

$$\begin{aligned}[P \times \omega_q]^{(q)} &= q^{-1}[P \times \omega_{q-1} \times \omega_1]^{(q)} \\ &= q^{-1}[P \times \omega_{q-1}]^{(q)} \times \omega_1 \\ &\qquad - 2(m-p-q+1)\,[P \times \omega_{q-1}]^{(q-1)} \\ &= q^{-1}\{[P \times \omega_{q-1}]^{(q-1)}\}^{(1)} \times \omega_1 \\ &\qquad - 2(m-p-q+1)\,[P \times \omega_{q-1}]^{(q-1)}.\end{aligned}$$

Hence, by the result for index $q-1$,

$$\begin{aligned}[P \times \omega_q]^{(q)} &= q^{-1}(-2)^{q-1}\frac{(m-p)\,!}{(m-p-q+1)\,!}\,P^{(1)} \times \omega_1 \\ &\qquad - 2(m-p-q+1)\,(-2)^{q-1}\frac{(m-p)\,!}{(m-p-q+1)\,!}\,P \\ &= (-2)^q\frac{(m-p)\,!}{(m-p-q)\,!}\,P, \quad \text{if } p+q \leqslant m,\end{aligned}$$

since, by hypothesis, $P^{(1)} = 0$; the same proof shows that

$$[P \times \omega_q]^{(q)} = 0, \quad \text{if } p+q > m.$$

42·3. Let us now consider any p-form $(0 \leqslant p \leqslant m)$, and let q be the integral part of $\tfrac{1}{2}p$, $q = [\tfrac{1}{2}p]$.

We define q p-forms $P_{(1)}, \ldots, P_{(q)}$ inductively, as follows:

$$P_{(1)} = P - \frac{(m-p+q)\,!}{(-2)^q\,(m-p+2q)\,!}\,P^{(q)} \times \omega_q,$$

$$P_{(r)} = P_{(r-1)} - \frac{(m-p+q-r+1)\,!}{(-2)^{q-r+1}(m-p+2q-2r+2)\,!}\,P_{(r-1)}^{(q-r+1)} \times \omega_{q-r+1},$$

where $P_{(r-1)}^{(q-r+1)}$ is a $(p-2q+2r-2)$-form defined from $P_{(r-1)}$ as in § 42·1. Then, by Lemma C,

$$P_{(1)}^{(q)} = 0.$$

We now prove that $\qquad P_{(r)}^{(q-r+1)} = 0.$

Since the result is true for $r = 1$, we assume it true for index

$r-1$ and prove it true for index r. By the hypothesis of induction,

$$[P^{(q-r+1)}_{(r-1)}]^{(1)} = P^{(q-r+2)}_{(r-1)} = 0.$$

Hence, by Lemma C,

$$P^{(q-r+1)}_{(r)} = P^{(q-r+1)}_{(r-1)} - P^{(q-r+1)}_{(r-1)}$$
$$= 0.$$

Write

$$Q_{p-2q+2r-2} = \frac{(m-p+q-r+1)!}{(-2)^{q-r+1}(m-p+2q-2r+2)!} P^{(q-r+1)}_{(r-1)}$$
$$(r=1, \ldots, q),$$
$$Q_p = P_{(p)},$$

Then
$$[Q_{p-2k}]^{(1)} = 0,$$

and
$$P = \sum_{k=0}^{q} Q_{p-2k} \times \omega_k.$$

We have thus found a convenient form in which to write any p-form on M $(0 \leqslant p \leqslant m)$.

42·4. We next find a convenient form when $p > m$. If P^* is the dual of P (§ 27·2), P^* is a form of multiplicity $2m-p<m$, and hence we can write

$$P^* = \sum_{k=0}^{q} R_{2m-p-2k} \times \omega_k, \qquad (q = [m - \tfrac{1}{2}p]),$$

where
$$[R_{2m-p-2k}]^{(1)} = 0.$$

Then
$$P = (-1)^p \sum_{k=0}^{q} [R_{2m-p-2k} \times \omega_k]^*.$$

It is therefore necessary to calculate the dual of a form

$$Q_{r-2k} \times \omega_k \qquad (r \leqslant m),$$

where
$$[Q_{r-2k}]^{(1)} = 0.$$

In order to do this, we use geodesic coordinates and consider a typical term of Q_{r-2k}, say,

$$Q_{i_1 \ldots i_a\, m+i_1 \ldots m+i_a\, j_1 \ldots j_b\, m+k_1 \ldots m+k_c}$$
$$\times\, dx^{i_1} \ldots dx^{i_a}\, dx^{m+i_1} \ldots dx^{m+i_a}\, dx^{j_1} \ldots dx^{j_b}\, dx^{m+k_1} \ldots dx^{m+k_c}$$

(in which the summation convention is not used). Here i_l, j_m, k_n are all different, and are not greater than m. The condition

$$[Q_{r-2k}]^{(1)} = 0$$

can be written as

$$\sum_{\alpha=1}^{m} Q_{\alpha\, i_2 \dots i_a\, m+\alpha\, m+i_2 \dots m+i_a\, j_1 \dots j_b\, m+k_1 \dots m+k_c} = 0.$$

Now $\quad \omega_k = (-1)^{\frac{1}{2}k(k-1)} \Sigma\, dx^{i_1} \dots dx^{i_k}\, dx^{m+i_1} \dots dx^{m+i_k};$

hence the typical term of $Q_{r-2k} \times \omega_k$ is

$$(-1)^{\frac{1}{2}k(k-1)+ak} \Sigma\, Q_{\alpha_1 \dots \alpha_a\, m+\alpha_1 \dots m+\alpha_a\, j_1 \dots j_b\, m+k_1 \dots m+k_c}$$
$$\times\, dx^{i_1} \dots dx^{i_{a+k}}\, dx^{m+i_1} \dots dx^{m+i_{a+k}}\, dx^{j_1} \dots dx^{j_b}\, dx^{m+k_1} \dots dx^{m+k_c},$$

where the summation is over the sets $(\alpha_1, \dots, \alpha_a)$ taken from $(i_1, \dots, i_{a+k})$. Using the condition

$$[Q_{r-2k}]^{(1)} = 0,$$

we can replace the summation by summation over the sets $(\alpha_1, \dots, \alpha_a)$ taken from the integers not greater than m which are different from $i_1, \dots, i_{a+k}, j_1, \dots, j_b, k_1, \dots, k_c$, provided we multiply by $(-1)^a$. To form the dual we have to replace the r differentials dx^{i_l}, dx^{m+i_l}, dx^{j_m}, dx^{m+k_n} by the remaining $2m-r$ differentials. The resulting term is seen at once to be a term of

$$Q'_{r-2k} \times \omega_{m-r+k},$$

save for sign. An examination of the permutations involved leads to the equation

$$[Q_{r-2k} \times \omega_k]^* = (-1)^{\epsilon+k}\, Q'_{r-2k} \times \omega_{m-r+k},$$

where $\quad \epsilon = \tfrac{1}{2}m(m-1) + \tfrac{1}{2}r(r+1).$

Hence if P is a p-form, $p > m$,

$$P = \sum_{0}^{q} [R_{2m-p-2k} \times \omega_k]^* \qquad q = [m - \tfrac{1}{2}p]$$
$$= (-1)^{\frac{1}{2}m(m+1)+\frac{1}{2}p(p-1)} \sum_{0}^{q} (-1)^k R'_{2m-p-2k} \times \omega_{p-m+k},$$

where $\quad [R'_{2m-p-2k}]^{(1)} = 0.$

The results for $p > m$ and $p \leqslant m$ can be summed up in the theorem:

THEOREM III. *Any p-form P on M may be written as*

$$P = \sum_{a,b} P_a \times \omega_b,$$

where $$a + 2b = p,$$

and the summation is over all non-negative integral values of a, b for which

$$a + b \leqslant m,$$

and where $$[P_a]^{(1)} = 0.$$

The dual of P is given by

$$P^* = (-1)^{\epsilon} \sum_{a,b} (-1)^b P'_a \times \omega_{m-p+b},$$

where $$\epsilon = \tfrac{1}{2}m(m-1) + \tfrac{1}{2}p(p+1).$$

42·5. The form given by Theorem III is unique. In order to prove this, we show that if

$$\sum_{a,b} P_a \times \omega_b = 0, \quad [P_a]^{(1)} = 0,$$

where $$a + 2b = p, \quad a + b \leqslant m,$$

then $$P_a = 0 \quad \text{(all permissible values of } a).$$

Suppose, indeed, that not all P_a are zero, and let a_1 be the least a for which $P_a \neq 0$. Then if

$$a_1 + 2b_1 = p,$$

$$\begin{aligned} 0 &= \sum_{a,b} [P_a \times \omega_b]^{(b_1)} \\ &= \sum_{a,b} [(P_a \times \omega_b)^{(b)}]^{(b_1 - b)} \\ &= \sum_{a,b} (-2)^b \frac{(m-a)!}{(m-a-b)!} P_a^{(b_1 - b)} \\ &= (-2)^{b_1} \frac{(m-a_1)!}{(m-a_1-b_1)!} P_{a_1}. \end{aligned}$$

Thus $$P_{a_1} = 0,$$

and we have a contradiction. The formula given by Theorem III is therefore unique.

42·6. Finally we obtain an alternative form for the condition

$$P^{(1)} = 0,$$

for any form P of multiplicity $p \leqslant m$. Suppose first that this condition is satisfied. Then, in the first place,

$$[P \times \omega_{m-p+1}]^{(1)} = 0,$$

by Lemma B. But, by Theorem III, we can write

$$P \times \omega_{m-p+1} = \sum_{a,b} P_a \times \omega_b,$$

where $$[P_a]^{(1)} = 0$$

and $$a+b \leqslant m,$$

$$a+2b = 2m-p+2.$$

Hence $$0 = \sum_{a,b} [P_a \times \omega_b]^{(1)}$$

$$= -2\sum_{a,b} (m-a-b+1)\, P_a \times \omega_{b-1},$$

by Lemma B. Hence $P_a = 0$, for each admissible a,

and therefore $$P \times \omega_{m-p+1} = 0.$$

Conversely, if $$P \times \omega_{m-p+1} = 0,$$

let $$P = \sum_{k=0}^{q} Q_{p-2k} \times \omega_k, \quad q = [\tfrac{1}{2}p],$$

where $$[Q_{p-2k}]^{(1)} = 0.$$

Then

$$P \times \omega_{m-p+1} = \sum_{k=0}^{q} \binom{m-p+k+1}{k} Q_{p-2k} \times \omega_{m-p+k+1} = 0.$$

Therefore, by Lemma B,

$$0 = [P \times \omega_{m-p+1}]^{(1)}$$

$$= -2\sum_{k=0}^{q} k \binom{m-p+k+1}{k} Q_{p-2k} \times \omega_{m-p+k}.$$

It follows from the uniqueness theorem proved in connection with Theorem III that

$$Q_{p-2k} = 0 \qquad (k = 1, 2, \ldots, q).$$

Therefore $$P = Q_p,$$

where $$[Q_p]^{(1)} = 0.$$

Hence the condition $$P^{(1)} = 0$$

can be replaced by $$P \times \omega_{m-p+1} = 0.$$

42·7. A similar argument shows that, if $p > m$, a p-form P satisfies the condition

$$P^{(1)} = 0$$

only if $$P = 0.$$

Indeed, by Theorem III, we can write

$$P = \sum_{a,b} P_a \times \omega_b \qquad (a + 2b = p),$$

summed for $a + b \leqslant m$. If

$$P^{(1)} = 0,$$

we have, by Lemma B,

$$0 = -2 \sum_{a,b} (m - a - b + 1) P_a \times \omega_{b-1},$$

and hence $$P_a = 0,$$

for all admissible a. Hence P is zero.

43·1. The classification of harmonic integrals on an algebraic manifold. With the aid of the results found in §§ 42·1–42·7 we arrive immediately at certain theorems of importance in the theory of harmonic integrals.

Theorem IV. *If P is a harmonic p-form, $Q = P \times \omega_1$ is a harmonic $(p+2)$-form, or else it is zero.*

In the first place,

$$\begin{aligned} Q_x &= P_x \times \omega_1 + (-1)^p P \times (\omega_1)_x \\ &= 0. \end{aligned}$$

Secondly,

$$g^{rs}Q_{i_1\ldots i_{p+1}r,s} = \frac{1}{p!\,2!}g^{rs}\delta^{j_1\ldots j_{p+2}}_{i_1\ldots i_{p+1}r}\,g_{m+j_1\,j_2}P_{j_3\ldots j_{p+2},s}$$

$$= 0,$$

since, P being harmonic,

$$g^{rs}P_{i_1\ldots i_{p-1}\,r,s} = 0,$$

and $$g^{r\,m+s}P_{i_1\ldots i_{p-1}\,r,s} = 0$$

(Theorem I, Corollary).

43·2. Now let us consider a harmonic p-form P, where $0 \leqslant p \leqslant m$. By Theorem III,

$$P = \sum_0^q Q_{p-2k} \times \omega_k \qquad (q=[\tfrac{1}{2}p]),$$

where $$[Q_{p-2k}]^{(1)} = 0.$$

If we turn to the construction of the forms Q_{p-2k}, we recall that, in the notation of § 42, we first constructed forms $P_{(1)}, \ldots, P_{(q)}$. Now,

$$Q_{p-2q} = \frac{(m-p+q)!}{(-2)^q\,(m-p+2q)!}P^{(q)}.$$

Since, by Theorem I, $P^{(q)}$ is harmonic, Q_{p-2q} is harmonic. It follows that $P_{(1)}$ is harmonic, and therefore

$$Q_{p-2q+2} = \frac{(m-p+q-1)!}{(-2)^{q-1}\,(m-p+2q-2)!}P^{(q-1)}_{(1)}$$

is harmonic. It follows by a simple induction that Q_{p-2k} is harmonic, for $k = 0, \ldots, q$.

Next, let us consider a p-form P, where $p > m$. Write it in the form, given in Theorem III,

$$P = \sum_{a,b} Q_a \times \omega_b \qquad (a+2b=p;\ a+b \leqslant m).$$

Then the dual form P^* is

$$P^* = \pm \sum_{a,b} (-1)^b Q'_a \times \omega_b,$$

and is harmonic. Since $2m-p<m$, and

$$[Q'_a]^{(1)} = 0,$$

it follows that Q'_a is harmonic. Therefore, by Theorem II, Q_a is harmonic.

If we now use Theorem IV, we obtain

THEOREM V. *When a harmonic p-form P is written in the form*

$$P = \sum_{a,b} Q_a \times \omega_b \qquad (a+2b=p;\ a+b \leqslant m),$$

as described in Theorem III,

$$Q_a \quad \text{and} \quad Q_a \times \omega_b$$

are harmonic, for all admissible values of a and b.

43·3. The harmonic forms P which satisfy the condition

$$P^{(1)} = 0$$

play an important role in the development of our theory, and it is convenient to give them a name. We shall call them *effective forms*, and their integrals will be called *effective integrals*.

Let P be any harmonic form of multiplicity $p \leqslant m$. Then, by Theorem V, we can write

$$P = \overline{P} + \sum_{k=1}^{q} P_{p-2k} \times \omega_k \qquad (q = [\tfrac{1}{2}p]),$$

where
$$Q = \sum_{k=1}^{q} k^{-1} P_{p-2k} \times \omega_{k-1}$$

is a harmonic $(p-2)$-form and $\overline{P}$ is an effective p-form. Hence we can write

$$P = \overline{P} + Q \times \omega_1.$$

Conversely, if Q is any harmonic $(p-2)$-form $(2 \leqslant p \leqslant m)$, and $\bar{P}$ is an effective p-form, we can write

$$Q = \sum_{k=1}^{q} k^{-1} P_{p-2k} \times \omega_{k-1},$$

where

$$[P_{p-2k}]^{(1)} = 0;$$

then

$$P = \bar{P} + Q \times \omega_1$$

$$= \sum_{0}^{q} P_{p-2k} \times \omega_k,$$

where $P_p = \bar{P}$, is an harmonic p-form written in the standard form. Now there are R_p harmonic p-forms, and R_{p-2} harmonic $(p-2)$-forms. It follows that there are exactly $R_p - R_{p-2}$ effective forms, provided that

$$\bar{P} + Q \times \omega_1$$

is not zero unless

$$\bar{P} = 0 \quad \text{and} \quad Q = 0.$$

Since $p \leqslant m$, the uniqueness theorem of § 42·5 shows that this condition is satisfied, and hence we have, by a simple induction:

THEOREM VI. *There are exactly $S_p = R_p - R_{p-2}$ independent effective integrals of multiplicity p $(0 \leqslant p \leqslant m)$. If we denote these by*

$$\int P_p^i \qquad (i = 1, \ldots, S_p),$$

a basis for the R_p harmonic integrals of multiplicity p is given by the $q+1$ sets of integrals

$$\int P_p^i \qquad (i = 1, \ldots, S_p),$$

$$\int P_{p-2}^i \times \omega_1 \qquad (i = 1, \ldots, S_{p-2}),$$

$$\vdots$$

$$\int P_{p-2q}^i \times \omega_q \qquad (i = 1, \ldots, S_{p-2q}),$$

where $q = [\frac{1}{2}p]$.

We call the integrals

$$\int P^i_{p-2h} \times \omega_h$$

the *ineffective integrals of the hth class*, and the forms appearing in the integrands the *ineffective forms of the hth class*. An ineffective integral (form) of the 0th class is the same as an effective integral (form).

We notice that the condition that P should be an effective p-form $(0 \leqslant p \leqslant m)$ can be written as

$$P \to 0,$$

$$P' \to 0,$$

$$P \times \omega_{m-p+1} = 0.$$

44. **Topology of algebraic manifolds.** Before we can discuss the periods of harmonic integrals on an algebraic manifold, we must recall the main topological properties of the manifold. The theorems which we require were either discovered by Lefschetz, or are simple deductions from his results; and the reader is referred to his tract [7], or to Zariski's summary [17], for the details of the proofs.

Lefschetz's theory of cycles on the manifold corresponding to a variety V_m provides a classification of cycles relative to a linear system of varieties V_{m-1} of dimension $m-1$, which may be taken as the system of prime sections of V_{m-1}. In constructing the Riemannian manifold of V_m in § 37·1, we took a particular representation of V_m as a non-singular variety in a projective space S_r, and constructed the Riemannian of S_r by Mannoury's method. We now take the system of varieties of dimension $m-1$ on V_m, relative to which we propose to classify the cycles of M, to be the system of prime sections of this representation of V_m, and we denote the section of V_m by a generic linear space of S_r of dimension $r-m+p$ by V_p $(0 \leqslant p \leqslant m)$. Corresponding to V_p there exists a sub-manifold of $2p$ dimensions on M, which we denote by $M(V_p)$.

The theorems which we shall require in the investigations of the periods of harmonic integrals are as follows. We need only consider homology with division.

I. If $p \leqslant m$, and Γ_q is any q-cycle of M $(q \leqslant p)$, there is a cycle lying in $M(V_p)$ homologous to Γ_q.

II. Every homology between the q-cycles of M lying in $M(V_p)$ $(p > q)$ which holds in M holds also in $M(V_p)$. Hence $R_q(V_p) = R_q$.

III. If $p \leqslant m$, there exists a basis Γ_p^i $(i = 1, \ldots, R_p(V_p))$ for the p-cycles of $M(V_p)$ in which the cycles fall into $q+2$ classes $(q = [\tfrac{1}{2}p])$:

$$\Gamma_p^i \quad (i = 1, \ldots, R_p - R_{p-2}),$$

$$\Gamma_p^i \quad (i = R_p - R_{p-2} + 1, \ldots, R_p - R_{p-4}),$$

$$\vdots$$

$$\Gamma_p^i \quad (i = R_p - R_{p-2q} + 1, \ldots, R_p),$$

$$\Gamma_p^i \quad (i = R_p + 1, \ldots, R_p(V_p)),$$

with the following properties.

(i) The cycles Γ_p^i $(i = 1, \ldots, R_p)$ form a basis for the p-cycles of M. We call them the *invariant cycles* of $M(V_p)$. The cycles Γ_p^i $(i = R_p + 1, \ldots, R_p(V_p))$, which we call the *vanishing cycles* of $M(V_p)$, are homologous to zero in M.

(ii) A cycle of the hth class has zero intersection (in $M(V_p)$) with any cycle of the kth class, if $h \neq k$. The cycles

$$\Gamma_p^i \quad (i = R_p - R_{p-2h} + 1, \ldots, R_p - R_{p-2h-2})$$

are called *ineffective cycles of the hth class* $(h = 0, \ldots, q)$. An ineffective cycle of the 0th class is also called an *effective cycle*.

(iii) The ineffective p-cycles of the hth class are homologous to cycles of $M(V_{p-h})$, but no linear combination Γ_p of them is homologous to a cycle in the general $M(V_{p-h-1})$. There may, however, exist an algebraic variety D of dimension less than $p-h$ on V_m with the property that $M(D)$ contains a cycle homologous to Γ_p. The variety D is not in this case a generic linear section of V_m.

IV. If $p<m$, in order to classify the $(2m-p)$-cycles of M we first classify the p-cycles of M according to the scheme of III. Then, a basis for the $(2m-p)$-cycles of M exists consisting of cycles Γ^i_{2m-p} $(i=1, \ldots, R_{2m-p}=R_p)$ which fall into $q+1$ classes $(q=[\frac{1}{2}p])$:

$$\Gamma^i_{2m-p} \quad (i=1, \ldots, R_p-R_{p-2}),$$
$$\Gamma^i_{2m-p} \quad (i=R_p-R_{p-2}+1, \ldots, R_p-R_{p-4}),$$
$$\vdots$$
$$\Gamma^i_{2m-p} \quad (i=R_p-R_{p-2q}+1, \ldots, R_p),$$

with the properties:

(i) $$\Gamma^i_{2m-p} \cdot M(V_p) \approx \Gamma^i_p;$$

(ii) $$(\Gamma^i_p \cdot \Gamma^j_{2m-p}) = 0$$
$$(R_p-R_{p-2h}<i\leqslant R_p-R_{p-2h-2};\ R_{p-2k}<j\leqslant R_p-R_{p-2k-2})$$

if $h\neq k$;

(iii) if $$R_p-R_{p-2h}<i\leqslant R_p-R_{p-2h-2},$$

then $$\Gamma^i_{2m-p} \cdot M(V_{p-h}) = \Gamma_{p-2h}$$

is an effective $(p-2h)$-cycle (see V, below).

Consider on V_{m-h} a linear ∞^{m-p+h} system of varieties V_{p-2h} with the property that through a general point of V_{m-h} there passes just one variety of the system. In $M(V_{p-2h})$ there is a cycle homologous to Γ_{p-2h}, and, as V_{p-2h} describes the system, this cycle describes a locus (in the sense of Lefschetz) which is a $(2m-p)$-cycle lying in $M(V_{m-h})$ and homologous to Γ^i_{2m-p}.

In order to preserve a continuity in our formulae it is convenient to call the cycles

$$\Gamma^i_{2m-p} \quad (R_p-R_{p-2h}<i\leqslant R_p-R_{p-2h-2})$$

ineffective cycles of class $2m-p-m+h=m-p+h$. We can then state the result, which includes III (iii):

V. If Γ_r is an r-cycle (any r) of class h, Γ_r is homologous to a cycle of $M(V_{r-h})$, and, further,

$$\Gamma_r \,.\, M(V_{m-k})$$

is an ineffective $(r-2k)$-cycle of class $h-k$ if $k \leqslant h$, and is homologous to zero if $k > h$.

VI. When the system $|V_{m-1}|$ of prime sections of V_m is given, the set of ineffective r-cycles of class h is uniquely determined. If another system $|U_{m-1}|$ on V_m which gives a projective representation U_m of V_m as a variety without singular points is chosen, the classification of the r-cycles may be altered. But the classification of the r-cycles into sets of ineffective cycles of different classes is unaltered in the special case when U_{m-1} is algebraically equivalent to V_{m-1}, and $|U_{m-1}|$ has no base points on V_m; hence it is fixed by the complete continuous system $\{V_{m-1}\}$ defined by the prime sections. Thus if we are given on V_m any complete continuous system $\{C\}$ of varieties of dimension $m-1$, which contains a linear system $|C|$ without base elements providing a projective representation of V_m as a variety without multiple points, there corresponds a unique classification of the cycles of V_m having the properties described in I–V.

45·1. Periods of harmonic integrals. The reader will have observed a correspondence between the results and terminology of §§ 43 and 44. We now show the precise nature of this correspondence. To do this we first require to prove the formula

$$\int_{\Gamma_{h+2k}} P \times \omega_k = (-1)^{mk+\frac{1}{2}k(k+1)} \frac{(4\pi)^k}{k!} \int_{\Gamma_h} P,$$

where $\qquad \Gamma_h \approx \Gamma_{h+2k} \,.\, M(V_{m-k}),$

and P is any closed h-form.

We suppose the cycles of M are classified as in § 44 and use the notation of that paragraph. In the notation of § 22·5, we have

$$\int_{\Gamma_{h+2k}} P \times \omega_k = \sum_{i,j} \lambda_{ij}^{2m-k}(\Gamma_{h+2k}) \int_{\Gamma_h^i} P \times \int_{\Gamma_{2k}^j} \omega_k.$$

But $$\int_{\Gamma^j_{2k}} \omega_k = (-1)^{\frac{1}{2}k(k-1)} \frac{(4\pi)^k}{k!} n\, \delta^j_{R_{2k}}$$

by § 41·1, where n is the order of V_m, and hence

$$\int_{\Gamma_{h+2k}} P \times \omega_k = (-1)^{\frac{1}{2}k(k-1)} \frac{(4\pi)^k}{k!} n \sum_i \lambda^{2m-2k}_{iR_{2k}}(\Gamma_{h+2k}) \int_{\Gamma^i_h} P.$$

Let $$\mathbf{a}_p = \| (\Gamma^i_p \, . \, \Gamma^j_{2m-p}) \|,$$

$$\mathbf{a}_{2m-2k}(\Gamma_{h+2k}) = \| (\Gamma_{h+2k} \, . \, \Gamma^i_{2m-2k} \, . \, \Gamma^j_{2m-h}) \| .$$

Then $$\mathbf{a}_{2k} = \begin{pmatrix} \mathbf{a}^0_{2k} & 0 & \dots & \\ 0 & \mathbf{a}^1_{2k} & & \\ \vdots & & \ddots & \\ & & & \mathbf{a}^k_{2k} \end{pmatrix},$$

where $\mathbf{a}^i_{2k}$ is a square matrix of $R_{2k-2i} - R_{2k-2i-2}$ rows and columns, and $\mathbf{a}^k_{2k} = (-1)^{k(m-k)} n$. Now

$$[\boldsymbol{\lambda}^{2m-2k}(\Gamma_{h+2k})]' = \mathbf{a}^{-1}_{2m-2k} \, . \, \mathbf{a}_{2m-2k}(\Gamma_{h+2k}) \, . \, \mathbf{a}^{-1}_h.$$

Therefore

$$\begin{aligned}[\boldsymbol{\lambda}^{2m-2k}(\Gamma_{h+2k})]_{iR_{2k}} &= [\mathbf{a}^{-1}_{2m-2k} \, . \, \mathbf{a}_{2m-2k}(\Gamma_{h+2k}) \, . \, \mathbf{a}^{-1}_h]_{R_{2k} i} \\ &= (-1)^{k(m-k)} n^{-1} [\mathbf{a}_{2m-2k}(\Gamma_{h+2k}) \, . \, \mathbf{a}^{-1}_h]_{R_{2k} i}.\end{aligned}$$

But if $$\Gamma_h \approx \mu_i \Gamma^i_h \approx \Gamma_{h+2k} \, . \, \Gamma^{R_{2k}}_{2m-2k},$$

then $$[\mathbf{a}_{2m-2k}(\Gamma_{h+2k})]_{R_{2k} j} = \sum_l \mu_l (\mathbf{a}_h)_{lj}.$$

Hence $$[\boldsymbol{\lambda}^{2m-2k}(\Gamma_{h+2k})]_{iR_{2k}} = (-1)^{k(m-k)} n^{-1} \mu_i.$$

Therefore,

$$\begin{aligned}\int_{\Gamma_{h+2k}} P \times \omega_k &= (-1)^{mk+\frac{1}{2}k(k+1)} \frac{(4\pi)^k}{k!} \sum_i \mu_i \int_{\Gamma^i_h} P \\ &= (-1)^{mk+\frac{1}{2}k(k+1)} \frac{(4\pi)^k}{k!} \int_{\Gamma_h} P \\ &= \nu_k \int_{\Gamma_h} P, \text{ say.}\end{aligned}$$

45·2. We use this result to show that, if $p \leqslant m$, an effective p-fold integral has non-zero periods only on the effective p-cycles. Let Γ_p be a p-cycle of class $h \geqslant 1$. Then there exists a $(2m-p+2)$-cycle Γ_{2m-p+2} of class $m-p+h+1$ with the property

$$\Gamma_{2m-p+2} \cdot M(V_{p-1}) \approx \Gamma_p.$$

If P is an effective p-form,

$$P \times \omega_{m-p+1} = 0.$$

Therefore,

$$\int_{\Gamma_p} P = (\nu_{m-p+1})^{-1} \int_{\Gamma_{2m-p+2}} P \times \omega_{m-p+1}$$
$$= 0.$$

In general, let P be an effective h-form. We may call $P \times \omega_k$ an ineffective $(h+2k)$-form of class k. This is in agreement with the definition given in § 43·3 for the cases $h+2k \leqslant p$. If Γ_{h+2k} is a cycle of class r, we have

$$\int_{\Gamma_{h+2k}} P \times \omega_k = \nu_k \int_{\Gamma_h} P,$$

where
$$\Gamma_h \approx \Gamma_{h+2k} \cdot M(V_{m-k})$$

is an ineffective cycle of class $r-k$. But

$$\int_{\Gamma_h} P$$

is zero unless $r = k$. Hence:

THEOREM VII. *An ineffective integral of multiplicity p and class k has non-zero periods only on the p-cycles of class k.* Since there are $S_{p-2k} = R_{p-2k} - R_{p-2k-2}$ ineffective p-fold integrals of class k, and S_{p-2k} ineffective cycles of class k, and since the period matrix of the R_p harmonic integrals of multiplicity p is non-singular, it follows that *the ineffective integrals of class k have a non-singular period matrix with respect to the ineffective cycles of class k.*

The results of this paragraph show that the periods of the harmonic integrals can be determined as soon as we know the periods of the effective integrals on the effective cycles. Moreover, since the dual of an effective p-form P $(p \leqslant m)$ is $\pm P' \times \omega_{m-p}$, where P' is an effective form, we may reduce the study of the periods of harmonic integrals and their relations with the periods of their duals, to the study of the periods of the integrals of the effective forms P and their relations with the periods of the integrals of the associated forms P'. This study can be confined to the sub-manifold $M(V_p)$.

46·1. Complex parameters. In view of the results proved in the preceding paragraph, we shall confine our considerations to the effective harmonic p-forms, and to the effective p-cycles, $1 \leqslant p \leqslant m$. Let

$$P = \frac{1}{p!} P_{i_1 \dots i_p} dx^{i_1} \dots dx^{i_p}$$

be an effective harmonic p-form, so that

$$P \times \omega_{m-p+1} = 0.$$

Then, by Theorem II of § 40,

$$P' = \frac{1}{p!} P_{m+i_1 \dots m+i_p} dx^{i_1} \dots dx^{i_p}$$

is a harmonic p-form, and

$$P' \times \omega_{m-p+1} = 0.$$

Let us express the forms in terms of the complex parameters

$$z_h = x_h + i x_{m+h},$$

and their conjugate imaginaries

$$\bar{z}_h = x_h - i x_{m+h}.$$

We have

$$P + (-i)^p P' = \frac{1}{p!} P_{i_1 \dots i_p} [dx^{i_1} \dots dx^{i_p} + i^p dx^{m+i_1} \dots dx^{m+i_p}].$$

Consider the typical term

$$dx^{i_1}\dots dx^{i_r}\,dx^{m+i_1}\dots dx^{m+i_r}[dx^{j_1}\dots dx^{j_s}\,dx^{m+k_1}\dots dx^{m+k_t}$$
$$+(-1)^{r+t}\,i^{s+t}\,dx^{m+j_1}\dots dx^{m+j_s}\,dx^{k_1}\dots dx^{k_t}]$$
$$=\frac{i^{r-t}}{2^{r+s+t}}dz^{i_1}\dots dz^{i_r}\,d\bar{z}^{i_1}\dots d\bar{z}^{i_r}\left[\prod_1^s(dz^{j_a}+d\bar{z}^{j_a})\prod_1^t(dz^{k_b}-d\bar{z}^{k_b})\right.$$
$$\left.+(-1)^r\prod_1^s(dz^{j_a}-d\bar{z}^{j_a})\prod_1^t(dz^{k_b}+d\bar{z}^{k_b})\right].$$

The right-hand side of this identity is a sum of terms each involving an even number of conjugate imaginary differentials $d\bar{z}^h$. Hence we can write†

$$P+(-i)^p\,P'=\sum_{k=0}^{q}P_{(2k)}\qquad (q=[\tfrac{1}{2}p]),$$

where

$$P_{(2k)}=[(p-2k)!\,(2k)!]^{-1}\,P^{(2k)}_{i_1\dots i_{p-2k};\,j_1\dots j_{2k}}$$
$$\times\, dz^{i_1}\dots dz^{i_{p-2k}}\,d\bar{z}^{j_1}\dots d\bar{z}^{j_{2k}},$$

where in applying the summation convention we sum for values of the indices from 1 to m, and $P^{(2k)}_{i_1\dots i_{p-2k};\,j_1\dots j_{2k}}$ is skew-symmetric in the suffices $i_1,\dots,i_{p-2k}$; $j_1,\dots,j_{2k}$, separately. Now

$$P+(-i)^p\,P'\to 0.$$

If we express this condition in terms of the complex parameters, we find that

$$\sum_{r=1}^{p-2k+1}\frac{\partial}{\partial z_{i_r}}P^{(2k)}_{i_1\dots i_{r-1}\,i_{r+1}\dots i_{p-2k+1};\,j_1\dots j_{2k}}=0,$$

$$\sum_{r=1}^{2k+1}(-1)^{r-1}\frac{\partial}{\partial \bar{z}_{j_r}}P^{(2k)}_{i_1\dots i_{p-2k};\,j_1\dots j_{r-1}\,j_{r+1}\dots j_{2k+1}}=0,$$

for each k. Hence $\qquad P_{(2k)}\to 0.$

† The suffix $2k$ is enclosed in brackets, in order to distinguish the form from the symbol often used to indicate a form of multiplicity $2k$.

Similarly, it may be shown that we can write

$$P-(-i)^p P' = \sum_{k=0}^{r} P_{(2k+1)} \qquad (r=[\tfrac{1}{2}(p-1)]),$$

where

$$P_{(2k+1)} \to 0,$$

and hence

$$2P = \sum_{k=0}^{p} P_{(k)},$$

where

$$P_{(k)} \to 0.$$

46·2. Next consider ω_1. At the origin of geodesic coordinates we have

$$\omega_1 = \sum_1^m dx^i\, dx^{m+i}$$

$$= \frac{i}{2}\sum_1^m dz^i\, d\bar{z}^i,$$

and hence in any set of parameters

$$\omega_1 = a_{i;\,j}\, dz^i\, d\bar{z}^j.$$

Similarly,

$$\omega_{m-p+1} = \frac{1}{[(m-p+1)\,!]^2}\, a_{i_1\ldots i_{m-p+1};\; j_1\ldots j_{m-p+1}} \times dz^{i_1}\ldots dz^{i_{m-p+1}}\, d\bar{z}^{j_1}\ldots d\bar{z}^{j_{m-p+1}}.$$

It follows immediately that

$$P\times\omega_{m-p+1}=0$$

implies

$$P_{(k)}\times\omega_{m-p+1}=0 \qquad (k=0,\ldots,p).$$

Conversely, let

$$P_{(k)} = [(p-k)\,!\,k\,!]^{-1}\, P^{(k)}_{i_1\ldots i_{p-k};\; j_1\ldots j_k}\, dz^{i_1}\ldots dz^{i_{p-k}}\, d\bar{z}^{j_1}\ldots d\bar{z}^{j_k}$$

be a closed form such that

$$P_{(k)}\times\omega_{m-p+1} = 0.$$

Let

$$z_h = x_h + ix_{m+h}, \quad \bar{z}_h = x_h - ix_{m+h},$$

and

$$P_{(k)} = A + iB.$$

Then an elementary calculation shows that, if $p = 2q$,

$$A_{m+i_1\dots m+i_p} = (-1)^{q+k} A_{i_1\dots i_p},$$

and, if $p = 2q-1$,

$$A_{m+i_1\dots m+i_p} = (-1)^{q+k} B_{i_1\dots i_p}.$$

It follows that if P is either the real or the imaginary part of $P_{(k)}$, then

$$P \to 0,$$

$$P \times \omega_{m-p+1} \to 0,$$

and
$$P' = \frac{1}{p!} P_{m+i_1\dots m+i_p} dx^{i_1} \dots dx^{i_p} \to 0.$$

Hence P is an effective p-form. We call $P_{(k)}$ an effective form *of type k*. Similarly a p-form which is the product of an effective $(p-2h)$-form of type k by ω_h is called an ineffective form *of type $h+k$*. We then have

THEOREM VIII. *An effective p-form P can be written as a sum*

$$P = \sum_{k=0}^{p} P_{(k)},$$

where $P_{(k)}$ is an effective form of type k, that is an effective form which can be written as

$$P_{(k)} = [(p-k)!\,k!]^{-1} P^{(k)}_{i_1\dots i_{p-k};\,j_1\dots j_k} dz^{i_1} \dots dz^{i_{p-k}} d\bar{z}^{j_1} \dots d\bar{z}^{j_k}.$$

46·3. Consider the closed form

$$P_{(0)} = \frac{1}{p!} P_{i_1\dots i_p} dz^{i_1} \dots dz^{i_p}.$$

The condition
$$P_{(0)} \times \omega_{m-p+1} = 0$$

imposes no restriction on $P_{(0)}$. Since $P_{(0)}$ is closed, we find that the coefficients of the form depend only on $(z_1, \dots, z_m)$, and not on $(\bar{z}_1, \dots, \bar{z}_m)$. But it is a classical theorem in the theory of algebraic functions and their integrals that an integral

$$\int \frac{1}{p!} A_{i_1\dots i_p} dz^{i_1} \dots dz^{i_p}$$

whose coefficients are analytic functions of the local parameters at every point of an algebraic variety, and which satisfies the equations

$$\sum_{r=1}^{p+1}(-1)^{r-1}\frac{\partial}{\partial z_{i_r}}A_{i_1\ldots i_{r-1}i_{r+1}\ldots i_{p+1}}=0,$$

is an algebraic integral of the first kind attached to the variety. If the variety is projected into a variety with only ordinary singularities in a space of $m+1$ dimensions, in which the non-homogeneous cartesian coordinates are $(z_1,\ldots,z_{m+1})$, and if the variety is given by

$$F(z_1,\ldots,z_{m+1})=0,$$

the integral can be written in the form

$$\int\frac{1}{p!}\frac{P_{i_1\ldots i_p}(z_1,\ldots,z_{m+1})}{\dfrac{\partial F}{\partial z_{m+1}}}dz^{i_1}\ldots dz^{i_p},$$

where $$P_{i_1\ldots i_p}(z_1,\ldots,z_{m+1})\qquad(i_r=1,\ldots,m)$$

is a polynomial in $(z_1,\ldots,z_{m+1})$, adjoint to $F(z_1,\ldots,z_{m+1})$, and satisfying certain other relations (cf. [4] for a complete statement of the conditions). Conversely, the real and imaginary parts of an algebraic integral of the first kind are effective integrals.

47·1. **Properties of the period matrices of effective integrals.** Let us consider the $S_p=R_p-R_{p-2}$ effective p-fold integrals, divided up into integrals of types h $(h=0,\ldots,p)$. Let the integrands be the forms

$$P^i_{(0)}\qquad(i=1,\ldots,\rho^0_p),$$
$$\vdots$$
$$P^i_{(p)}\qquad(i=1,\ldots,\rho^p_p),$$

where the forms $P^i_{(h)}$ $(i = 1, \ldots, \rho^h_p)$ are effective forms of type h, and we may assume

$$P^i_{(h)} = \bar{P}^i_{(p-h)}.$$

We have
$$\sum_{h=0}^{p} \rho^h_p = S_p, \quad \rho^h_p = \rho^{p-h}_p.$$

Let the period matrix of the integrals of the type h be $\mathbf{\Omega}^p_h$, a matrix of ρ^h_p rows and S_p columns. We may regard the period cycles of the integrals as cycles in $M(V_p)$. The matrix of their intersections in this sub-manifold is non-singular. We denote the transpose of its inverse by $\mathbf{A}^p_0$.

By the bilinear relations (§ 22·4), we have the equations

$$\mathbf{\Omega}^p_h \mathbf{A}^p_0 (\bar{\mathbf{\Omega}}^p_k)' = \left\| \int_{M(V_p)} P^i_{(h)} \times \bar{P}^j_{(k)} \right\|.$$

(*a*) If $h \neq k$, $P^i_{(h)} \times \bar{P}^j_{(k)} = 0$ in $M(V_p)$, since the product is a $2p$-form with $p-h+k$ differentials dz^i, and $p+h-k$ differentials $d\bar{z}^i$. Hence

$$\mathbf{\Omega}^p_h \mathbf{A}^p_0 (\bar{\mathbf{\Omega}}^p_k)' = 0 \qquad (h \neq k).$$

(*b*) Let $h = k$. If we write

$$\mathbf{\Omega}^p_h \mathbf{A}^p_0 (\bar{\mathbf{\Omega}}^p_h)' = i^p(-1)^{p+h} \boldsymbol{\alpha}_h,$$

we can prove that $\boldsymbol{\alpha}_h$ is a positive definite Hermitian matrix. To do this, let $\lambda_1, \ldots, \lambda_{\rho^h_p}$ be any complex numbers, not all zero. Then

$$\lambda_i \alpha^{ij}_h \bar{\lambda}_j = i^p(-1)^h \sum_{i,j} \lambda_i [\mathbf{\Omega}^p_h \mathbf{A}^p_0 (\bar{\mathbf{\Omega}}^p_h)']_{ij} \bar{\lambda}_j$$

$$= i^p(-1)^h \int_{M(V_p)} P \times \bar{P},$$

where
$$P = \lambda_i P^i_{(h)}$$

is an effective integral of type h, different from zero. We have to show that the right-hand side of this equation is positive.

Let A, iB be the real and imaginary parts of P, so that

$$P = A+iB.$$

If $p = 2q$,
$$\begin{aligned} P\times\bar{P} &= (A+iB)\times(A-iB) \\ &= A\times A+B\times B \\ &= (-1)^{q+h}[A\times A'+B\times B'], \end{aligned}$$

by the result of § 46·2; and, if $p = 2q-1$,

$$\begin{aligned} P\times\bar{P} &= (A+iB)\times(A-iB) \\ &= -2iA\times B \\ &= -2i(-1)^{q+h}A\times A' \\ &= -i(-1)^{q+h}[A\times A'+B\times B']. \end{aligned}$$

Hence, in either case,

$$i^p(-1)^h\int_{M(V_p)} P\times\bar{P} = (-1)^p\int_{M(V_p)}[A\times A'+B\times B'].$$

Now, by the results of §§ 45·1 and 42·4, we have

$$\begin{aligned} \int_{M(V_p)} A\times A' &= \xi\int_M A\times A'\times\omega_{m-p} \\ &= \xi\eta\int_M A\times A^*, \end{aligned}$$

A^* being the dual of A, and similarly,

$$\int_{M(V_p)} B\times B' = \xi\eta\int_M B\times B^*,$$

where
$$\xi = (-1)^{m(m-p)+\frac{1}{2}(m-p)(m-p+1)}\frac{(m-p)!}{(4\pi)^{m-p}},$$

and
$$\eta = (-1)^{\frac{1}{2}m(m-1)+\frac{1}{2}p(p+1)}.$$

Hence
$$\xi\eta = (-1)^p\frac{(m-p)!}{(4\pi)^{m-p}},$$

and therefore we have

$$i^p(-1)^h\int_{M(V_p)} P\times\bar{P} = \frac{(m-p)!}{(4\pi)^{m-p}}\int_M[A\times A^*+B\times B^*] > 0,$$

since A and B are not both zero.

47·2. We can obtain similar results for the ineffective p-fold integrals of class h. An ineffective p-form P of class h is the product of an effective $(p-2h)$-form by ω_h. Now the effective $(p-2h)$-forms of class h can be arranged into $p-2h+1$ types, those of type k being of the form

$$Q = [(p-2h-k)!\,k!]^{-1}\,Q_{i_1\dots i_{p-2h-k};\,j_1\dots j_k}\,dz^{i_1}\dots dz^{i_{p-2h-k}}d\bar{z}^{j_1}\dots d\bar{z}^{j_k}.$$

Recalling the form of ω_h found in § 46·2, we see that

$$\begin{aligned} P &= Q\times\omega_h \\ &= [(p-h-k)!\,(h+k)!]^{-1}\,P_{i_1\dots i_{p-h-k};\,j_1\dots j_k} \\ &\qquad\qquad \times dz^{i_1}\dots dz^{i_{p-h-k}}d\bar{z}^{j_1}\dots d\bar{z}^{j_{h+k}}, \end{aligned}$$

which we have called an ineffective form of type $h+k$. Thus the ineffective forms of class h can be arranged into sets of type k, where $k = h, \dots, p-h$.

Let $\mathbf{\Lambda}^p_a$ be the period matrix of the ineffective p-fold integrals of class h and type a with respect to the ineffective cycles of class h of M, which may be taken as cycles of $M(V_p)$. Let $\mathbf{A}^p_h$ be the transpose of the inverse of the intersection matrix of these cycles. Then, as before,

$$\mathbf{\Lambda}^p_a\,\mathbf{A}^p_h(\bar{\mathbf{\Lambda}}^p_b)' = \mathbf{0} \qquad (a \neq b).$$

We now show that

$$\mathbf{\Lambda}^p_a\,\mathbf{A}^p_h(\bar{\mathbf{\Lambda}}^p_a)' = i^p(-1)^{p+h+a}\,\boldsymbol{\beta}_a,$$

where $\boldsymbol{\beta}_a$ is a positive definite Hermitian matrix. Following the argument given above, we have to show that if P is any ineffective form of class h and type a

$$i^p(-1)^{h+a}\int_{M(V_p)} P\times\bar{P} > 0.$$

Now $P = Q\times\omega_h$, where Q is an effective $(p-2h)$-form of type $a-h$. Hence

$$\begin{aligned} \int_{M(V_p)} P\times\bar{P} &= \frac{(2h)!}{(h!)^2}\int_{M(V_p)} Q\times\bar{Q}\times\omega_{2h} \\ &= \frac{(4\pi)^{2h}}{(h!)^2}(-1)^h\int_{M(V_{p-2h})} Q\times\bar{Q} \\ &= \frac{(4\pi)^{2h}}{(h!)^2}(-1)^h\,(i)^{p-2h}\,(-1)^{p+a-h}\,k, \end{aligned}$$

where k is a real number greater than zero. Hence

$$i^p(-1)^{h+a}\int_{M(V_p)} P\times\bar{P}>0.$$

It will be observed that if we consider all the harmonic forms of type h, including the effective and ineffective integrals,

$$\mathbf{\Omega}\mathbf{A}\bar{\mathbf{\Omega}}' = i^p(-1)^h\boldsymbol{\gamma}_h,$$

where $\mathbf{\Omega}$ is the period matrix of integrals, $\mathbf{A}$ is the transpose of the inverse of the intersection matrix of the period cycles, and $\boldsymbol{\gamma}_h$ is a Hermitian matrix, which is not, however, positive definite.

47·3. Any closed p-form which can be written in the form

$$[(p-k)!\,k!]^{-1}P_{i_1\ldots i_{p-k};\,j_1\ldots j_k}dz^{i_1}\ldots dz^{i_{p-k}}\,d\bar{z}^{j_1}\ldots d\bar{z}^{j_k},$$

whether it is harmonic or not, will be called a form of type k. We now prove a lemma concerning a set of p-forms of type h whose integrals have zero periods on all ineffective cycles of class j ($j=0,\ldots,k-1,\,k+1,\ldots,[\tfrac{1}{2}p]$). Let

$$Q^1,\ldots,Q^\rho$$

be such a set of forms. If we choose as basis for the p-cycles of M the cycles Γ^i_p ($i=1,\ldots,R_p$) of § 44, the integral of Q^i has zero periods on all the cycles except

$$\Gamma^i_p\quad(R_p-R_{p-2k}<i\leqslant R_p-R_{p-2k-2}),$$

and there exists an ineffective form of class k whose integral has the same periods. Then

$$Q^i\sim\sum_a\lambda^i_{(a)j}P^j_{(a)},$$

where the forms $P^j_{(a)}$ are ineffective forms of class k and type a. Let $\mathbf{\Omega}$ be the period matrix of the integrals of $Q^1,\ldots,Q^\rho$ and $\mathbf{\Lambda}^p_a$ the period matrix of the integrals of $P^j_{(a)}$ ($j=1,2,\ldots$). If $\mathbf{A}^p_k$ be the transpose of the inverse of the intersection matrix of the ineffective cycles of class k, calculated in $M(V_p)$, then by calculations similar to those made above,

$$\mathbf{\Omega A}_k^p(\bar{\mathbf{\Lambda}}_a^p)' = \mathbf{0} \qquad (a \neq h).$$

But we have
$$\mathbf{\Omega} = \sum_a \mathbf{\lambda}_{(a)} \mathbf{\Lambda}_a^p.$$

Since
$$\mathbf{\Lambda}_b^p \mathbf{A}_k^p(\bar{\mathbf{\Lambda}}_a^p)' = \mathbf{0} \qquad (a \neq b),$$

we have
$$\mathbf{\lambda}_{(a)} \mathbf{\Lambda}_a^p \mathbf{A}_h^p(\bar{\mathbf{\Lambda}}_a^p)' = \mathbf{0} \qquad (a \neq h).$$

Since $\mathbf{\Lambda}_a^p \mathbf{A}_h^p(\bar{\mathbf{\Lambda}}_a^p)'$ is non-singular,

$$\mathbf{\lambda}_{(a)} = \mathbf{0} \qquad (a \neq h).$$

Further,
$$\begin{aligned}\mathbf{\Omega A}_h^p \bar{\mathbf{\Omega}}' &= \mathbf{\lambda}_{(h)} \mathbf{\Lambda}_h^p \mathbf{A}_h^p(\bar{\mathbf{\Lambda}}_h^p)' (\bar{\mathbf{\lambda}}_{(h)})' \\ &= i^p(-1)^{p+h+k}\mathbf{M},\end{aligned}$$

where $\mathbf{M}$ is a positive definite or indefinite matrix according as there does not, or does, exist a linear combination of the forms Q_i which is null.

47·4. We make two deductions from this result. We confine ourselves for simplicity to integrals having zero periods on the ineffective cycles, but the generalisation is immediate.

(i) Let $\mathbf{\Omega}_h^p$ be the period matrix of the effective p-fold integrals of V_m of type h, and let $\tilde{\mathbf{\Omega}}_h^p$ be the period matrix of the effective p-fold integrals of V_k $(m > k > p)$, the metric on $M(V_k)$ being determined from the Mannoury representation of V_k obtained from the representation of V_m. Since the effective p-cycles of $M(V_k)$ coincide with the effective p-cycles of M, the effective p-fold integrals of type h on V_m define integrals of type h on $M(V_k)$ satisfying the conditions of our lemma. Hence

$$\mathbf{\Omega}_h^p = \mathbf{\lambda}_{(h)} \tilde{\mathbf{\Omega}}_h^p.$$

Hence
$$\mathbf{\Omega}^p = \begin{pmatrix} \mathbf{\Omega}_0^p \\ \vdots \\ \mathbf{\Omega}_p^p \end{pmatrix} = \begin{pmatrix} \mathbf{\lambda}_{(0)} & & \\ & \ddots & \\ & & \mathbf{\lambda}_{(p)} \end{pmatrix} \begin{pmatrix} \tilde{\mathbf{\Omega}}_0^p \\ \vdots \\ \tilde{\mathbf{\Omega}}_p^p \end{pmatrix} = \mathbf{\Lambda} \tilde{\mathbf{\Omega}}^p.$$

Now $\mathbf{\Omega}^p$ and $\tilde{\mathbf{\Omega}}^p$ are non-singular matrices of $R_p - R_{p-2}$ rows and columns. Hence $\mathbf{\Lambda}$ is a non-singular matrix. This is only possible if $\rho_p^h = \rho_p^h(V_k)$, and $\boldsymbol{\lambda}_{(h)}$ is a non-singular matrix of ρ_p^h rows and columns. By rearranging the integrals, we can write

$$\boldsymbol{\Omega}_h^p = \tilde{\boldsymbol{\Omega}}_h^p.$$

(ii) If $k = p$, the same reasoning shows that

$$\boldsymbol{\Omega}_h^p = \boldsymbol{\lambda}_{(h)} \tilde{\boldsymbol{\Omega}}_h^p,$$

where $\boldsymbol{\lambda}_{(h)}$ is a matrix of ρ_p^h rows and $\rho_p^h(V_p)$ rows and columns, since the effective p-cycles of M are included amongst the effective p-cycles of $M(V_p)$.

48. **Change of metric.** A third application of the lemma of § 47·3 enables us to establish the result that the matrices $\boldsymbol{\Omega}_h^p$ are birational invariants of M relative to the complete continuous system defined by the prime sections of M. It will be recalled that the classification of the p-cycles of M into classes of ineffective cycles given in § 44 is unaltered if we replace V_m by a birationally equivalent variety U_m which is (a) in (1-1) correspondence with V_m without exception, (b) such that the prime sections of U_m correspond to varieties of $(m-1)$ dimensions on V_m belonging to the complete continuous system defined by the prime sections on V_m. The harmonic integrals, on the other hand, depend on the choice of metric on M.

If
$$P_{(h)}^i \qquad (i = 1, \dots, \rho_p^h)$$
are the effective p-fold forms on M of type h $(h = 0, \dots, p)$ and
$$\tilde{P}_{(h)}^i \qquad (i = 1, \dots, \tilde{\rho}_p^h)$$
are the forms defined on M by the effective integrals of type h $(h = 0, \dots, p)$ defined by a Mannoury metric associated with U_m, the forms
$$\tilde{P}_{(h)}^i \qquad (i = 1, \dots, \tilde{\rho}_p^h)$$
satisfy the conditions of the lemma of § 47·3. If we calculate the periods of all the integrals with respect to the same basis for the effective p-cycles, the lemma gives the result

$$\tilde{\boldsymbol{\Omega}}_h^p = \boldsymbol{\lambda}_{(h)} \boldsymbol{\Omega}_h^p,$$

where $\boldsymbol{\lambda}_{(h)}$ is a matrix of $\tilde{\rho}_p^h$ rows and ρ_p^h columns. Hence

$$\begin{pmatrix} \tilde{\boldsymbol{\Omega}}_0^p \\ \tilde{\boldsymbol{\Omega}}_1^p \\ \vdots \\ \tilde{\boldsymbol{\Omega}}_p^p \end{pmatrix} = \begin{pmatrix} \boldsymbol{\lambda}_{(0)} & \mathbf{0} & \cdots & \\ \mathbf{0} & \boldsymbol{\lambda}_{(1)} & & \\ \vdots & & \ddots & \\ & & & \boldsymbol{\lambda}_{(p)} \end{pmatrix} \begin{pmatrix} \boldsymbol{\Omega}_0^p \\ \boldsymbol{\Omega}_1^p \\ \vdots \\ \boldsymbol{\Omega}_p^p \end{pmatrix}.$$

But the two period matrices in this equation are non-singular matrices of $R_p - R_{p-2}$ rows and columns. Hence

$$\begin{pmatrix} \boldsymbol{\lambda}_{(0)} & \mathbf{0} & \cdots & \\ \mathbf{0} & \boldsymbol{\lambda}_{(1)} & & \\ \vdots & & \ddots & \\ & & & \boldsymbol{\lambda}_{(p)} \end{pmatrix}$$

is non-singular, and therefore $\tilde{\rho}_p^h = \rho_p^h$ and $\boldsymbol{\lambda}_{(h)}$ is non-singular. By rearranging the integrals, we can write

$$\tilde{\boldsymbol{\Omega}}_h^p = \boldsymbol{\Omega}_h^p.$$

In general, if the periods are calculated with respect to arbitrary (possibly different) bases for the effective cycles, we have

$$\tilde{\boldsymbol{\Omega}}_h^p = \boldsymbol{\lambda}_{(h)} \boldsymbol{\Omega}_h^p \mathbf{A},$$

where $\boldsymbol{\lambda}_{(h)}$ is a non-singular square matrix, and $\mathbf{A}$ is a non-singular square matrix of integers. Period matrices which satisfy an equation of this form are said to be equivalent. Hence we have the theorem:

On an algebraic variety of m dimensions we consider a complete continuous system of varieties of $m-1$ dimensions $\{C\}$, and in $\{C\}$ we take two linear systems $|C_1|$ and $|C_2|$ with the same base points, which serve to give projective representations of the variety as varieties without singularities. If we now proceed to build up the theory of harmonic integrals as in the earlier part of this chapter, the period matrices of the effective p-fold integrals of type h on the two varieties are equivalent.

An exactly similar argument holds when we consider in-effective integrals of class k instead of effective integrals.

49. Some enumerative results. Certain formulae concerning the numbers ρ_p^h can now be proved. We have

$$\rho_p^h = \rho_p^{p-h},$$

and

$$\sum_{k=0}^{p} \rho_p^k = R_p - R_{p-2}.$$

Let us first suppose that p is odd, say $p = 2q+1$. Then

$$2\sum_{k=0}^{q} \rho_p^k = R_p - R_{p-2}.$$

If $p = 0, 1, R_{p-2} = 0$, and hence, by a simple induction, we find that R_p is even whenever p is odd. This result has been proved otherwise by Lefschetz [7].

Next let us suppose that p is even, say $p = 2q$. Then, by the result of § 47·1,

$$\mathbf{\Omega}_h^p \mathbf{A}_0^p (\bar{\mathbf{\Omega}}_k^p)' = \mathbf{0} \qquad (h \neq k),$$

$$\mathbf{\Omega}_h^p \mathbf{A}_0^p (\bar{\mathbf{\Omega}}_h^p)' = (-1)^{q+h} \boldsymbol{\alpha}_h,$$

where $\mathbf{A}_0^p$ is the inverse of the transpose of the intersection matrix of the effective p-cycles (regarded as cycles in $M(V_p)$), and $\boldsymbol{\alpha}_h$ is a positive definite Hermitian matrix. $\mathbf{A}_0^p$ is a real symmetric matrix, and it has a character called the *signature*, defined as follows. Let $\mathbf{T}$ be a non-singular matrix such that $\mathbf{T}\mathbf{A}_0^p\bar{\mathbf{T}}'$ is a real diagonal matrix. The matrix $\mathbf{T}$ is not uniquely defined by this property, but it is known [3] that the number of elements of $\mathbf{T}\mathbf{A}_0^p\bar{\mathbf{T}}'$ which are positive is independent of the matrix $\mathbf{T}$. This number is called the signature of $\mathbf{A}_0^p$. Clearly it is also the signature of $\mathbf{a}_p$, the intersection matrix of the effective p-cycles. We denote it by $S(\mathbf{A}_0^p)$ or $S(\mathbf{a}_p)$, and it is a relative invariant of V_m. Since $\boldsymbol{\alpha}_h$ is positive definite, there exists a matrix $\boldsymbol{\beta}_h$ such that

$$\boldsymbol{\beta}_h \bar{\boldsymbol{\beta}}_h' = \boldsymbol{\alpha}_h.$$

Take

$$\mathbf{T} = \begin{pmatrix} \boldsymbol{\beta}_0 & \mathbf{0} & \cdots & \\ \mathbf{0} & \boldsymbol{\beta}_1 & & \\ \vdots & & \ddots & \\ & & & \boldsymbol{\beta}_p \end{pmatrix}^{-1} \begin{pmatrix} \mathbf{\Omega}_0^p \\ \mathbf{\Omega}_1^p \\ \vdots \\ \mathbf{\Omega}_p^p \end{pmatrix}.$$

Then
$$\mathbf{T}\mathbf{A}^p\overline{\mathbf{T}}' = (-1)^q \begin{pmatrix} \boldsymbol{\epsilon}_0 & \cdots & & \\ \vdots & \boldsymbol{\epsilon}_1 & & \\ & & \ddots & \\ & & & \boldsymbol{\epsilon}_p \end{pmatrix},$$

where $\boldsymbol{\epsilon}_r$ is $(-1)^r$ times the unit matrix of ρ_p^r rows and columns. Hence
$$S(\mathbf{a}_p) = \rho_p^0 + \rho_p^1 + \ldots + \rho_p^p \qquad (q \text{ even})$$

and
$$S(\mathbf{a}_p) = \rho_p^1 + \rho_p^3 + \ldots + \rho_p^{p-1} \qquad (q \text{ odd}).$$

The number ρ_p^0 of everywhere finite p-fold algebraic integrals attached to V_m is an absolute invariant of V_m. We therefore have the following relations:

(i) $$\rho_1^0 = \tfrac{1}{2}R_1;$$

(ii) $$\rho_p^0 \leqslant \tfrac{1}{2}(R_p - R_{p-2})$$

when p is odd;

(iii) when $p = 2q$, $$\rho_{2q}^0 \leqslant S(\mathbf{a}_{2q}) \qquad (q \text{ even}),$$

$$\rho_{2q}^0 \leqslant R_{2q} - R_{2q-2} - S(\mathbf{a}_{2q}) \qquad (q \text{ odd}).$$

It is well known [2] that ρ_1^0 is equal to the first irregularity of V_m, and that ρ_m^0 is the geometric genus of V_m. For $m = 1$ we are concerned with an algebraic curve V_1, and the results obtained are classical in the theory of algebraic functions. In the case of surfaces ($m = 2$) we shall later obtain more precise results.

50·1. **Defective systems of integrals.** In the following paragraphs we prove a theorem which is a generalisation of Poincaré's theorem on defective integrals [11], for sets of integrals which have zero periods on all the ineffective cycles of M. A similar theorem can be proved for sets of integrals having non-zero periods only on the ineffective cycles of class k, for any fixed value of k, but the theorem cannot, as far as we know, be extended to sets of integrals which have non-zero periods on ineffective cycles of more than one class.

Let

$$\begin{array}{l} P^1_{(0)}, \ldots, P^{\sigma_0}_{(0)}, \\ P^1_{(1)}, \ldots, P^{\sigma_1}_{(1)}, \\ \vdots \\ P^1_{(p)}, \ldots, P^{\sigma_p}_{(p)} \end{array}$$

be a set of closed p-forms which have zero periods on all the ineffective p-cycles of M, with the properties:

(i) $P^i_{(h)}$ is a form of type h, i.e. it can be written as

$$P^i_{(h)} = [(p-h)!\,h!]^{-1} P_{i_1 \ldots i_{p-h};\, j_1 \ldots j_h} dz^{i_1} \ldots dz^{i_{p-h}} d\bar{z}^{j_1} \ldots d\bar{z}^{j_h},$$

and

$$P^i_{(h)} = \bar{P}^i_{(p-h)};$$

(ii) no linear combination of the forms with constant coefficients is null;

(iii) if $\mathbf{\Omega}_h$ is the period matrix of the integrals of the forms $P^i_{(h)}$ of type h, there exists a matrix $\mathbf{\Lambda}_h$ of σ_h rows and $\sigma = \sum_0^p \sigma_h$ columns, and a matrix $\mathbf{R}$ of integers of σ rows and $R_p - R_{p-2}$ columns such that

$$\mathbf{\Omega}_h = \mathbf{\Lambda}_h \mathbf{R}. \tag{5}$$

We call such a set of forms a *complete set*.

We first show that $\mathbf{\Lambda}_h$ is of rank σ_h, and $\mathbf{R}$ of rank σ. The $(p+1)$ equations (5) can be written together as

$$\mathbf{\Omega} = \begin{pmatrix} \mathbf{\Omega}_0 \\ \vdots \\ \mathbf{\Omega}_p \end{pmatrix} = \begin{pmatrix} \mathbf{\Lambda}_0 \\ \vdots \\ \mathbf{\Lambda}_p \end{pmatrix} \mathbf{R} = \mathbf{\Lambda}\mathbf{R}.$$

The matrix $\mathbf{\Omega}$ is of rank σ, otherwise there would be a linear combination of the forms $P^i_{(h)}$ which is null. Hence $\mathbf{\Lambda}$ and $\mathbf{R}$ are of rank σ at least. But $\mathbf{\Lambda}$ is a matrix of σ rows and columns, and $\mathbf{R}$ has σ rows; hence the rank of these matrices cannot exceed σ. If any $\mathbf{\Lambda}_h$ was of rank less than σ_h, $\mathbf{\Lambda}$ would be of rank less than σ. Hence $\mathbf{\Lambda}_h$ is of rank σ_h, and $\mathbf{R}$ is of rank σ.

Since $\mathbf{R}$ is of rank σ, there therefore exists a matrix $\mathbf{S}$ of integers of $R_p - R_{p-2} - \sigma$ rows and $R_p - R_{p-2}$ columns, so that

$$\mathbf{T} = \begin{pmatrix} \mathbf{R} \\ \mathbf{S} \end{pmatrix}$$

has determinant t different from zero. Let us make the change of base for the period cycles given by

$$t\Gamma_p^i \approx t_{ji}\Gamma_p'^j.$$

The integrals of our set of forms of type h have now period matrix equal to

$$t\mathbf{\Lambda}_h \mathbf{R}\mathbf{T}^{-1} = t\mathbf{\Lambda}_h(\mathbf{I}_\sigma, \mathbf{0}),$$

where $\mathbf{I}_\sigma$ is the unit matrix of σ rows and columns. The integrals of all the forms have therefore periods different from zero only on the cycles $\Gamma_p'^i$ $(i = 1, \ldots, \sigma)$. The matrices $\mathbf{\Omega}_h$ are now to be regarded as period matrices calculated for the cycles $\Gamma_p'^i$.

Let the intersection matrix of the effective cycles $\Gamma_p'^i$ $(i = 1, \ldots, R_p - R_{p-2})$ in $M(V_p)$ be written as

$$\begin{pmatrix} \mathbf{a}_1 & \mathbf{a}_2 \\ \mathbf{a}_3 & \mathbf{a}_4 \end{pmatrix},$$

where

$$\begin{aligned} \mathbf{a}_1 &= \| (\Gamma_p'^i \,.\, \Gamma_p'^j) \| && (i, j \leqslant \sigma), \\ \mathbf{a}_2 &= \| (\Gamma_p'^i \,.\, \Gamma_p'^j) \| && (i \leqslant \sigma; j > \sigma), \\ \mathbf{a}_3 &= \| (\Gamma_p'^i \,.\, \Gamma_p'^j) \| && (i > \sigma; j \leqslant \sigma), \\ \mathbf{a}_4 &= \| (\Gamma_p'^i \,.\, \Gamma_p'^j) \| && (i, j > \sigma), \end{aligned}$$

and write the transpose of its inverse similarly as

$$\begin{pmatrix} \mathbf{A}_1 & \mathbf{A}_2 \\ \mathbf{A}_3 & \mathbf{A}_4 \end{pmatrix},$$

where $\mathbf{A}_1$ has σ rows and columns, etc.

By Jacobi's theorem on inverse determinants we know that

$$|\mathbf{a}_4| = |\mathbf{A}_1|\, d,$$

where d is the determinant of the intersection matrix. We

prove that $|\mathbf{a}_4|$ is different from zero by showing that $\mathbf{A}_1$ is non-singular. By § 47·3, we have

$$\begin{pmatrix}\mathbf{\Omega}_0\\ \vdots \\ \mathbf{\Omega}_p\end{pmatrix}\begin{pmatrix}\mathbf{A}_1 & \mathbf{A}_2\\ \mathbf{A}_3 & \mathbf{A}_4\end{pmatrix}(\bar{\mathbf{\Omega}}'_0 \dots \bar{\mathbf{\Omega}}'_p) = (-i)^p \begin{pmatrix}\boldsymbol{\alpha}_0 & \cdots & \\ \vdots & \ddots & \\ & & \boldsymbol{\alpha}_p\end{pmatrix},$$

where $(-1)^h \boldsymbol{\alpha}_h$ is a positive definite Hermitian matrix. Hence

$$t^2\begin{pmatrix}\mathbf{\Lambda}_0\\ \vdots \\ \mathbf{\Lambda}_p\end{pmatrix}\mathbf{A}_1(\bar{\mathbf{\Lambda}}'_0 \dots \bar{\mathbf{\Lambda}}'_p) = (-i)^p \begin{pmatrix}\boldsymbol{\alpha}_0 & \cdots & \\ \vdots & \ddots & \\ & & \boldsymbol{\alpha}_p\end{pmatrix},$$

and, since the period matrices in this equation are square matrices of rank σ, and the right-hand side is non-singular, $\mathbf{A}_1$ is non-singular. Therefore $\mathbf{a}_4$ is non-singular. Let

$$\boldsymbol{\lambda} = |\mathbf{a}_4|\,\mathbf{a}_2\mathbf{a}_4^{-1},$$

and make the change of base given by

$$\Delta_p^i = |\mathbf{a}_4|\,\Gamma_p'^i - \lambda_j^i \Gamma_p'^{\sigma+j} \qquad (i \leqslant \sigma),$$
$$\Delta_p^i = |\mathbf{a}_4|\,\Gamma_p'^i \qquad (i > \sigma).$$

The integrals of our forms have zero periods on the cycles Δ_p^i $(i > \sigma)$ and the intersection matrix of the $R_p - R_{p-2}$ effective cycles Δ_p^i is of the form

$$\begin{pmatrix}\mathbf{b}_1 & 0\\ 0 & \mathbf{b}_4\end{pmatrix},$$

where $\mathbf{b}_1$ is a (σ, σ) matrix of rank σ, and $\mathbf{b}_4$ is a non-singular matrix of $R_p - R_{p-2} - \sigma$ rows and columns. Thus a complete set of σ p-fold integrals which have periods only on the effective cycles has the property that *we can choose a base for the effective p-cycles Δ_p^i $(i = 1, \dots, R_p - R_{p-2})$ so that*

(*a*) *the integrals have non-zero periods only on the cycles Δ_p^i $(i \leqslant \sigma)$;*

(*b*) $$(\Delta_p^i \,.\, \Delta_p^j) = 0 \qquad (i \leqslant \sigma; j > \sigma);$$

(*c*) $$\|(\Delta_p^i \,.\, \Delta_p^j)\| \text{ is non-singular} \qquad (i, j \leqslant \sigma).$$

For brevity we shall refer to the cycles Δ_p^i $(i=1, \ldots, \sigma)$, or any linear combinations of them, as the period cycles of the complete set. We observe that the whole set of $R_p - R_{p-2}$ effective integrals forms a complete set, their period cycles being the effective cycles of M.

50·2. Let us now suppose that a complete set of σ p-forms

$$P_{(0)}^i \qquad (i = 1, \ldots, \sigma_0),$$
$$P_{(1)}^i \qquad (i = 1, \ldots, \sigma_1),$$
$$\vdots$$
$$P_{(p)}^i \qquad (i = 1, \ldots, \sigma_p)$$

can be arranged so that the forms

$$P_{(0)}^i \qquad (i = 1, \ldots, \lambda_0),$$
$$P_{(1}^i \qquad (i = 1, \ldots, \lambda_1),$$
$$\vdots$$
$$P_{(p)}^i \qquad (i = 1, \ldots, \lambda_p)$$

form a complete set, where

$$0 < \lambda = \sum_{h=0}^{p} \lambda_h < \sigma.$$

We first choose as a basis for the effective p-cycles of M a set of cycles in $M(V_p)$,

$$\Delta_p^i \qquad (i = 1, \ldots, R_p - R_{p-2}),$$

satisfying the condition of the theorem of § 50·1. We now consider only the period cycles $\Delta_p^1, \ldots, \Delta_p^\sigma$, and repeat the argument of § 50·1, replacing σ and $R_p - R_{p-2}$ by λ and σ respectively. We find a basis

$$\Gamma_p^i \qquad (i = 1, \ldots, R_p - R_{p-2}),$$

where
$$\Gamma_p^i = \Delta_p^i \qquad (i > \sigma),$$

for the effective p-cycles of M, so that the period matrix of the integrals of $P^i_{(h)}$ $(i=1, \ldots, \sigma_h)$ with respect to the cycles Γ^i_p $(i=1, \ldots, \sigma)$ is of the form

$$\boldsymbol{\Lambda}_h = \begin{pmatrix} \boldsymbol{\omega}^1_h & \mathbf{0} \\ \boldsymbol{\omega}^2_h & \boldsymbol{\omega}^3_h \end{pmatrix}.$$

Here $\boldsymbol{\omega}^1_h$ is of λ_h rows and λ columns, etc., and the intersection matrix of the cycles Γ^i_p $(i=1, \ldots, \sigma)$ is of the form

$$\begin{pmatrix} \mathbf{a}_1 & \mathbf{0} \\ \mathbf{0} & \mathbf{a}_2 \end{pmatrix},$$

where $\mathbf{a}_1$ is a non-singular matrix of λ rows and columns.

Let us now consider the relations of § 47·3. We obtain

$$\begin{pmatrix} \boldsymbol{\omega}^1_h & \mathbf{0} \\ \boldsymbol{\omega}^2_h & \boldsymbol{\omega}^3_h \end{pmatrix} \begin{pmatrix} \mathbf{A}_1 & \mathbf{0} \\ \mathbf{0} & \mathbf{A}_2 \end{pmatrix} \begin{pmatrix} (\bar{\boldsymbol{\omega}}^1_k)' & (\bar{\boldsymbol{\omega}}^2_k)' \\ \mathbf{0} & (\bar{\boldsymbol{\omega}}^3_k)' \end{pmatrix} = 0 \qquad (h \neq k),$$

where $\mathbf{A}_i$ is the transpose of the inverse of $\mathbf{a}_i$. Hence

$$\boldsymbol{\omega}^1_h \mathbf{A}_1 (\bar{\boldsymbol{\omega}}^1_k)' = 0,$$

$$\boldsymbol{\omega}^2_h \mathbf{A}_1 (\bar{\boldsymbol{\omega}}^1_k)' = 0 \quad (h \neq k).$$

The matrix

$$\boldsymbol{\psi}_h = \mathbf{A}_1 \| (\bar{\boldsymbol{\omega}}^1_0)' \ldots (\bar{\boldsymbol{\omega}}^1_{h-1})' (\bar{\boldsymbol{\omega}}^1_{h+1})' \ldots (\bar{\boldsymbol{\omega}}^1_p)' \|$$

is of rank $\lambda - \lambda_h$, and hence if $\mathbf{x}$ is any matrix of rank λ_h, of λ_h rows and λ columns, which satisfies

$$\mathbf{x}\boldsymbol{\psi}_h = 0,$$

any solution of the equation

$$\mathbf{y}\boldsymbol{\psi}_h = 0$$

is given by $\mathbf{y} = \boldsymbol{\alpha}\mathbf{x}.$

But we can take $\mathbf{x} = \boldsymbol{\omega}^1_h,$

and hence $\boldsymbol{\omega}^2_h = \boldsymbol{\alpha}_h \boldsymbol{\omega}^1_h,$

where $\boldsymbol{\alpha}_h$ is a matrix of $\sigma_h - \lambda_h$ rows and λ_h columns. Clearly, we have

$$\bar{\boldsymbol{\alpha}}_h = \boldsymbol{\alpha}_{p-h}.$$

If we now replace the σ_h forms

$$P^i_{(h)} \qquad (i = 1, \ldots, \sigma_h)$$

by the forms

$$Q^i_{(h)} \qquad (i = 1, \ldots, \sigma_h),$$

where

$$Q^i_{(h)} = P^i_{(h)} \qquad (i \leqslant \lambda_h),$$

$$Q^{\lambda_h+i}_{(h)} = P^{\lambda_h+i}_{(h)} - \alpha^i_{(h)j} P^j_{(h)},$$

the period matrix of the integrals of the forms $Q^i_{(h)}$ is of the form

$$\begin{pmatrix} \boldsymbol{\omega}^1_h & 0 \\ 0 & \boldsymbol{\omega}^3_h \end{pmatrix},$$

and hence the forms $Q^i_{(h)}$ $(i = \lambda_h + 1, \ldots, \sigma_h;\ h = 0, \ldots, p)$ comprise a complete set. The complete set $P^i_{(h)}$ $(i = 1, \ldots, \sigma_h;\ h = 0, \ldots, p)$ is said to be *defective*, and the two complete sub-sets are called *complementary complete sub-sets*. A complete set which is not defective is said to be *pure*. The period matrix

$$\boldsymbol{\Omega} = \begin{pmatrix} \boldsymbol{\Omega}_0 \\ \vdots \\ \boldsymbol{\Omega}_p \end{pmatrix}$$

of a complete set which is pure is called a *pure* period matrix, and the period matrix of a defective complete set is said to be *irreducible*, or *impure*. The period matrix of a complete set with respect to its period cycles is always a non-singular square matrix.

Let

$$\boldsymbol{\Omega} = \begin{pmatrix} \boldsymbol{\Omega}_0 \\ \vdots \\ \boldsymbol{\Omega}_p \end{pmatrix} \quad \text{and} \quad \boldsymbol{\Lambda} = \begin{pmatrix} \boldsymbol{\Lambda}_0 \\ \vdots \\ \boldsymbol{\Lambda}_p \end{pmatrix}$$

be period matrices of two complete sets of p-fold integrals, written in the usual form. Let $\boldsymbol{\Omega}_h$ have ρ_h rows and ρ columns, and $\boldsymbol{\Lambda}_h$ have σ_h rows and σ columns. Then

$$\rho = \sum_{h=0}^{p} \rho_h, \quad \sigma = \sum_{h=0}^{p} \sigma_h.$$

The matrices $\mathbf{\Omega}$ and $\mathbf{\Lambda}$ are said to be *equivalent* if

(1) $$\rho_h = \sigma_h \qquad (h = 0, \ldots, p);$$

(2) the matrices $\mathbf{\Omega}_h$, $\mathbf{\Lambda}_h$ satisfy equations

$$\boldsymbol{\alpha}_h \mathbf{\Omega}_h \mathbf{A} = \mathbf{\Lambda}_h,$$

where $\boldsymbol{\alpha}_h$ is a matrix, of ρ_h rows and columns, which is of rank ρ_h, and $\mathbf{A}$ is a non-singular matrix of integers, having ρ rows and columns.

By a finite number of repetitions of the argument given above, we prove immediately that the period matrix of a complete set of integrals is equivalent to

$$\begin{pmatrix} \mathbf{\Omega}_0^{(1)} & \cdots & & \\ \vdots & \mathbf{\Omega}_0^{(2)} & & \\ & & \ddots & \\ & & & \mathbf{\Omega}_0^{(s)} \\ & & \ddots & \\ & & & \ddots \\ \mathbf{\Omega}_p^{(1)} & \cdots & & \\ \vdots & \mathbf{\Omega}_p^{(2)} & & \\ & & \ddots & \\ & & & \mathbf{\Omega}_p^{(s)} \end{pmatrix},$$

where
$$\begin{pmatrix} \mathbf{\Omega}_0^{(i)} \\ \vdots \\ \mathbf{\Omega}_p^{(i)} \end{pmatrix} = \mathbf{\Omega}^{(i)}$$

is a pure period matrix. The integrals can also be arranged so that their period matrix is

$$\begin{pmatrix} \mathbf{\Omega}^{(1)} & \cdots & & \\ \vdots & \mathbf{\Omega}^{(2)} & & \\ & & \ddots & \\ & & & \mathbf{\Omega}^{(s)} \end{pmatrix}.$$

50·3. We now show that the reduced form of the period matrix of a complete set of integrals is, essentially, unique.

Suppose, indeed, that the period matrix of a complete set of σ forms can be reduced to

$$\mathbf{\Omega} = \begin{pmatrix} \mathbf{\Omega}^{(1)} & \dots & \\ \vdots & \ddots & \\ & & \mathbf{\Omega}^{(r)} \end{pmatrix},$$

where $\mathbf{\Omega}^{(i)}$ is a pure period matrix having period cycles

$$\Gamma_p^{r_1+\dots+r_{i-1}+1}, \dots, \Gamma_p^{r_1+\dots+r_i},$$

and suppose that it can also be reduced to the form

$$\mathbf{\Lambda} = \begin{pmatrix} \mathbf{\Lambda}^{(1)} & \dots & \\ \vdots & \ddots & \\ & & \mathbf{\Lambda}^{(s)} \end{pmatrix},$$

where $\mathbf{\Lambda}^{(i)}$ is a pure period matrix having period cycles

$$\Delta_p^{s_1+\dots+s_{i-1}+1}, \dots, \Delta_p^{s_1+\dots+s_i}.$$

If $\qquad \Gamma_p^i = c_{ji}\Delta_p^j \qquad (i = 1, \dots, \sigma = \Sigma r_i = \Sigma s_i),$

there exists an equation

$$\boldsymbol{\alpha}\mathbf{\Omega} = \mathbf{\Lambda}\mathbf{C},$$

where $\boldsymbol{\alpha}$ is a non-singular matrix of σ rows and columns and $\mathbf{C} = (c_{ij})$. We write this as

$$\begin{pmatrix} \boldsymbol{\alpha}^{11} & \dots & \boldsymbol{\alpha}^{1r} \\ \vdots & & \\ \boldsymbol{\alpha}^{s1} & \dots & \boldsymbol{\alpha}^{sr} \end{pmatrix} \begin{pmatrix} \mathbf{\Omega}^{(1)} & \dots & \\ \vdots & \ddots & \\ & & \mathbf{\Omega}^{(r)} \end{pmatrix} = \begin{pmatrix} \mathbf{\Lambda}^{(1)} & \dots & \\ \vdots & \ddots & \\ & & \mathbf{\Lambda}^{(s)} \end{pmatrix} \begin{pmatrix} \mathbf{C}^{11} & \dots & \mathbf{C}^{1r} \\ \vdots & & \\ \mathbf{C}^{s1} & \dots & \mathbf{C}^{sr} \end{pmatrix},$$

where each of the elements is a matrix and

$$\boldsymbol{\alpha}^{ij} = \begin{pmatrix} \boldsymbol{\alpha}^{ij}_{(0)} & \mathbf{0} & \dots & \\ \mathbf{0} & \boldsymbol{\alpha}^{ij}_{(1)} & & \\ \vdots & & \ddots & \\ & & & \boldsymbol{\alpha}^{ij}_{(p)} \end{pmatrix}$$

has s_i rows and r_j columns. The matrix equation implies

$$\boldsymbol{\alpha}^{ij}\mathbf{\Omega}^{(j)} = \mathbf{\Lambda}^{(i)}\mathbf{C}^{ij},$$

that is, $\qquad \boldsymbol{\alpha}^{ij}_{(h)}\mathbf{\Omega}^{(j)}_h = \mathbf{\Lambda}^{(i)}_h\mathbf{C}^{ij} \qquad (h = 0, \dots, p),$

when we consider the integrals of different types separately. Let r_j^h, s_i^h be the numbers of rows in $\mathbf{\Omega}_h^{(j)}$, $\mathbf{\Lambda}_h^{(i)}$, that is, the ranks of these matrices. Since $\boldsymbol{\alpha}_{(h)}^{ij}$ has s_i^h rows and r_j^h columns, its rank ρ_h^{ij} satisfies the inequality

$$\rho_h^{ij} \leqslant \min(s_i^h, r_j^h).$$

Now, $\mathbf{\Omega}^{(j)}$ and $\mathbf{\Lambda}^{(i)}$ are non-singular square matrices. Hence the matrices $\boldsymbol{\alpha}^{ij}$ and $\mathbf{C}^{ij}$ have the same rank, ρ^{ij}, say.

There exists a non-singular matrix $\mathbf{P}_{(h)}^{ij}$ of s_i^h rows and columns such that

$$\mathbf{P}_{(h)}^{ij}\boldsymbol{\alpha}_{(h)}^{ij} = \begin{pmatrix}\boldsymbol{\beta}_{(h)}^{ij}\\ \mathbf{0}\end{pmatrix},$$

where $\boldsymbol{\beta}_{(h)}^{ij}$ is a matrix with ρ_h^{ij} rows. There also exists a unimodular matrix of integers $\mathbf{D}^{ij}$ so that

$$\mathbf{D}^{ij}\mathbf{C}^{ij} = \begin{pmatrix}\mathbf{B}^{ij}\\ \mathbf{0}\end{pmatrix},$$

where $\mathbf{B}^{ij}$ has ρ^{ij} rows. The equation

$$\boldsymbol{\alpha}_{(h)}^{ij}\mathbf{\Omega}_h^{(j)} = \mathbf{\Lambda}_h^{(i)}\mathbf{C}^{ij}$$

then gives

$$\boldsymbol{\beta}_{(h)}^{ij}\mathbf{\Omega}_h^{(j)} = \mathbf{L}_h^{(i)}\mathbf{B}^{ij},$$

where $\mathbf{L}_h^{(i)}$ is the matrix of the elements in the first ρ_h^{ij} rows and ρ^{ij} columns of $\mathbf{P}_{(h)}^{ij}\mathbf{\Lambda}_h^{(i)}(\mathbf{D}^{ij})^{-1}$. This equation implies that the matrix $\mathbf{\Omega}^{(j)}$ cannot be a pure period matrix unless either $\rho_h^{ij} = 0$ $(h=0, \ldots, p)$, or $\rho_h^{ij} = r_j^h$ $(h=0, \ldots, p)$.

We consider the second case. In this case we obviously have $\rho^{ij} = r_j$. Moreover, since

$$\rho_h^{ij} \leqslant \min(s_i^h, r_j^h),$$

we have $s_i^h \geqslant r_j^h$. Also $\mathbf{B}^{ij}$ is not singular. The equation

$$\boldsymbol{\beta}_{(h)}^{ij}\mathbf{\Omega}_h^{(j)} = \mathbf{L}_h^{(i)}\mathbf{B}^{ij}$$

can be written as

$$\boldsymbol{\beta}_{(h)}^{ij}\mathbf{\Omega}_h^{(j)} = (\mathbf{I}_{r_j^h}, \mathbf{0})\,\mathbf{P}_{(h)}^{ij}\mathbf{\Lambda}_h^{(i)}(\mathbf{D}^{ij})^{-1}\,\mathbf{B}^{ij},$$

where $\mathbf{I}_\lambda$ is the unit matrix of λ rows and columns. Hence, since $\boldsymbol{\beta}^{ij}_{(h)}$ is non-singular, we obtain the equation

$$\boldsymbol{\gamma}^{ij}_{(h)}\boldsymbol{\Lambda}^{(i)}_h = \boldsymbol{\Omega}^{(j)}_h\mathbf{E}^{ij},$$

where $$\boldsymbol{\gamma}^{ij}_{(h)} = (\boldsymbol{\beta}^{ij}_{(h)})^{-1}(\mathbf{I}_{r^h_j}, \mathbf{0})\,\mathbf{P}^{ij}_{(h)},$$

and $$\mathbf{E}^{ij} = (\mathbf{B}^{ij})^{-1}\mathbf{D}^{ij}.$$

In the same way, this shows that $\boldsymbol{\Lambda}^{(i)}$ would be impure, unless $\rho^{ij}_h = s^h_i$ $(h = 0, \ldots, p)$. Hence we conclude that if $r_j \neq s_i$, $\boldsymbol{\alpha}^{ij} = 0$. If $r_j = s_i$, $\boldsymbol{\alpha}^{ij}$ is either zero or a non-singular square matrix, and in the latter case $\boldsymbol{\Omega}^{(j)}$, $\boldsymbol{\Lambda}^{(i)}$ are equivalent period matrices.

Now suppose that $\boldsymbol{\Omega}^{(1)}, \boldsymbol{\Omega}^{(2)}, \ldots, \boldsymbol{\Omega}^{(a)}$ are equivalent period matrices, and that $\boldsymbol{\Omega}^{(i)}$ $(i > a)$ is not similar to $\boldsymbol{\Omega}^{(1)}$. From the result just proved, we know that $\boldsymbol{\Omega}^{(1)}$ is similar to one $\boldsymbol{\Lambda}^{(i)}$ at least. Let it be similar to $\boldsymbol{\Lambda}^{(1)}, \ldots, \boldsymbol{\Lambda}^{(b)}$, but not to $\boldsymbol{\Lambda}^{(j)}$ $(j > b)$. Then in the matrix $\boldsymbol{\alpha}$ the elements in the first $ar_1 = as_1$ columns and the last $\sigma - br_1$ rows are zero, and the elements in the first br_1 rows and last $\sigma - ar_1$ columns are zero. But $\boldsymbol{\alpha}$ is non-singular. This is only possible if $a = b$. Therefore we conclude that the reduced forms $\boldsymbol{\Omega}$, $\boldsymbol{\Lambda}$ of the period matrix of the σ forms have the properties:

(i) $$r = s;$$

(ii) for a suitable arrangement of $\boldsymbol{\Lambda}^{(1)}, \ldots, \boldsymbol{\Lambda}^{(r)}$, the period matrices $\boldsymbol{\Omega}^{(i)}$ and $\boldsymbol{\Lambda}^{(i)}$ are equivalent $(i = 1, \ldots, r)$. Thus the reduced form of the period matrix is essentially unique.

On the other hand, the pure complete sets of integrals whose period matrices $\boldsymbol{\Omega}^{(i)}$ go to make up the period matrix $\boldsymbol{\Omega}$ need not be uniquely defined. For if $\boldsymbol{\Omega}^{(1)}$ is similar to $\boldsymbol{\Omega}^{(2)}$, there may exist an equivalent reduced period matrix $\boldsymbol{\Lambda}$ in which $\boldsymbol{\Lambda}^{(1)}$ and $\boldsymbol{\Lambda}^{(2)}$ are similar to $\boldsymbol{\Omega}^{(1)}$ and $\boldsymbol{\Omega}^{(2)}$, and in which $\boldsymbol{\alpha}^{12}$, $\boldsymbol{\alpha}^{21}$ are not zero matrices. Thus, if $\boldsymbol{\Omega}^{(1)} = \boldsymbol{\Omega}^{(2)}$ and $\boldsymbol{\Omega}^{(1)}$ is the period matrix of the integrals of

$$P^i_{(h)} \qquad (i = 1, \ldots, r^h_1;\ h = 0, \ldots, p),$$

and $\boldsymbol{\Omega}^{(2)}$ is the period matrix of the integrals of

$$Q^{(i)}_h \qquad (i = 1, \ldots, r^h_1;\ h = 0, \ldots, p),$$

the forms $$aP^i_{(h)} + bQ^i_{(h)} \qquad (i = 1, \ldots, r^h_1;\ h = 0, \ldots, p),$$

where a and b are integers, form a pure complete set, whose period matrix may be taken as one of the component period matrices of another reduction $\mathbf{\Lambda}$ of $\mathbf{\Omega}$.

50·4. An important example of defective integrals arises in connection with the p-fold effective integrals on M, when $p<m$. When we consider these as integrals on $M(V_p)$, we have a complete set of integrals on this sub-manifold. In general, they are not effective integrals on $M(V_p)$, but the results of § 47·4 shows that if

$$P_{(h)} = [(p-h)!\,h!]^{-1}\,P^{(h)}_{i_1\ldots i_{p-h};\,j_1\ldots j_h}\,dz^{i_1}\ldots dz^{i_{p-h}}\,d\bar{z}^{j_1}\ldots d\bar{z}^{j_h}$$

is an effective p-form of type h on M, there exists an effective form

$$Q_{(h)} = [(p-h)!\,h!]^{-1}\,P^{(h)}_{i_1\ldots i_{p-h};\,j_1\ldots j_h}\,dz^{i_1}\ldots dz^{i_{p-h}}\,d\bar{z}^{j_1}\ldots d\bar{z}^{j_h},$$

of type h on $M(V_p)$ such that the integrals of $P_{(h)}$ and $Q_{(h)}$ have the same periods on the effective p-cycles of $M(V_p)$. The $R_p - R_{p-2}$ forms $Q_{(h)}$ $(h=0,\ldots,p)$ so obtained therefore comprise a complete set, which is a sub-set of the complete set of effective forms on $M(V_p)$. There therefore exists a complementary complete set of effective forms on $M(V_p)$ whose period cycles are the vanishing cycles of $M(V_p)$.

51·1.† **Applications to problems in algebraic geometry.** The applications which have been made of the theory of harmonic integrals to problems in classical algebraic geometry are somewhat scattered. In the following paragraphs we give a brief account of some immediate applications, but where an extensive knowledge of the geometrical theory of algebraic varieties is necessary to the understanding of applications, we merely give references. In a later paragraph we shall give an account of some applications to the theory

† As explained in the preface, the following paragraphs merely summarize, for the benefit of those interested in algebraic geometry, the more important applications which have been made of the results of this chapter.

of algebraic surfaces, but we first consider some general problems.

We have already pointed out that a p-cycle of the manifold M is always homologous to a p-cycle lying in the sub-manifold which we have denoted by $M(V_p)$. The ineffective p-cycles of class h $(h \geqslant 1)$ are homologous to cycles lying in the sub-manifold $M(V_{p-1})$. In addition, an effective p-cycle may be homologous to a cycle lying in a sub-manifold $M(D)$, where D is an algebraic variety on V_m of dimension less than p. In this case, D does not lie in a generic variety V_{p-1}, and usually it possesses some special geometrical properties. Let Γ_p be any p-cycle of M. We say that Γ_p is *of rank* k if there exists an algebraic variety D on V_m, which may be reducible, of dimension $p-k$, such that Γ_p is homologous to a cycle of $M(D)$. The rank of a p-cycle clearly cannot exceed the integral part of $\frac{1}{2}p$. If $p = 2q$, and Γ_p is of rank q, we say that Γ_p is *algebraic*.

We now prove that if

$$P^i_{(h)} \qquad (i=1, \ldots, \tau_h;\ h=0, \ldots, p)$$

is a base for the harmonic forms on M, both effective and ineffective, where $P^i_{(h)}$ is of type h and $\overline{P}^i_{(h)} = P^i_{(p-h)}$, then necessary conditions that a p-cycle Γ_p be of rank k are

$$\int_{\Gamma_p} P^i_{(h)} = 0 \qquad (i=1, \ldots, \tau_h;\ h=0, \ldots, k-1).$$

It is clear that these conditions imply the further conditions

$$\int_{\Gamma_p} P^i_{(h)} = 0 \qquad (i=1, \ldots, \tau_h;\ h=p-k+1, \ldots, p).$$

If Γ_p is of rank k, it is homologous to a cycle of $M(D)$, where D is an algebraic variety of dimension $p-k$, lying on V_m. If D is reducible, we write it as $D = \sum_1^r D_i$, where $D_1, \ldots, D_r$ are its irreducible components. Each D_i is of dimension

$p-k$, at most. On $M(D_i)$ there exists a cycle Γ_p^i such that $\Gamma_p \approx \lambda_i \Gamma_p^i$, on M. It is sufficient to prove that

$$\int_{\Gamma_p^i} P^j_{(h)} = 0 \qquad (j=0, \ldots, \tau_h;\ h=0, \ldots, k-1),$$

for each value of i.

Let the dimension of D_i be s_i. Then D_i is given, locally, by a set of equations

$$z_h = f_h(u_1, \ldots, u_{s_i}) \qquad (h=1, \ldots, m),$$
$$\bar{z}_h = \bar{f}_h(\bar{u}_1, \ldots, \bar{u}_{s_i}) \qquad (h=1, \ldots, m).$$

Since $s_i \leqslant p-k$, we have, on D_i, if $h<k$,

$$P^j_{(h)} = [(p-h)!\,h!]^{-1} P_{i_1 \ldots i_{p-h};\, j_1 \ldots j_h} \frac{\partial z_{i_1}}{\partial u_{l_1}} \cdots \frac{\partial z_{i_{p-h}}}{\partial u_{l_{p-h}}}$$
$$\times \frac{\partial \bar{z}_{j_1}}{\partial \bar{u}_{m_1}} \cdots \frac{\partial \bar{z}_{j_h}}{\partial \bar{u}_{m_h}} du^{l_1} \ldots du^{l_{p-h}}\, d\bar{u}^{m_1} \ldots d\bar{u}^{m_h}$$
$$= 0,$$

since $$p-h>p-k \geqslant s_i,$$

and therefore $$\int_{\Gamma_p^i} P^j_{(h)} = 0 \qquad (h<k).$$

51·2. We have therefore found a set of necessary conditions to be satisfied in order that the cycle Γ_p be of rank k. The question whether they are sufficient is of great importance in the application of the theory of harmonic integrals to algebraic geometry, but as yet it can only be answered in special cases. Lefschetz [7] has proved that, in the case $m=p=2$, the 2-cycle Γ_2 is algebraic if and only if

$$\int_{\Gamma_2} Q = 0,$$

for all algebraic integrals $\int Q$ of the first kind on M. This is equivalent to our condition in the only case which arises when $m=2$.

We now consider the case $p = 2$, for any m. If the 2-cycle Γ_2 is not homologous to zero (with division), it can be of rank *one* at most. We now show that if each algebraic double integral of the first kind on M has zero period on Γ_2, there exists an integer λ, different from zero, such that $\lambda\Gamma_2$ is algebraic. Except for the introduction of the multiplier λ, this is the same as proving the sufficiency of the conditions found in § 51·1.

If Γ_2 is any 2-cycle of M, there exists a non-zero integer λ such that $\lambda\Gamma_2$ is homologous to an invariant cycle Γ'_2 of $M(V_2)$. The cycle Γ'_2 is algebraic if the harmonic integrals of multiplicity *two* on $M(V_2)$,

$$\int Q^i_{(0)},$$

which are of type *zero* have zero periods on it. We saw, however, in § 50·4, that the harmonic integrals on $M(V_2)$ form a defective set, and that they can be decomposed into complementary complete sets, one having periods only on the invariant cycles and the other having periods only on the vanishing cycles. The necessary and sufficient condition that Γ'_2 be algebraic is that the algebraic double integrals (i.e. the harmonic integrals of type *zero*) on $M(V_2)$ which have zero periods on the vanishing cycles should have zero periods on Γ'_2. But the algebraic integrals in question are homologous (and indeed equal) to the integrals on $M(V_2)$ derived from the algebraic integrals on M (by § 47·4). Hence $\lambda\Gamma_2$ is algebraic if

$$\int_{\lambda\Gamma_2} P^i_{(0)} = \lambda \int_{\Gamma_2} P^i_{(0)} = 0,$$

for each effective integral of multiplicity *two* and type *zero* on M.

Lefschetz has shown [7] that the necessary and sufficient condition that a $(2m-2)$-cycle Γ_{2m-2} be algebraic is that the 2-cycle $\Gamma_2 = \Gamma_{2m-2} \cdot M(V_2)$ be algebraic. The condition for this is

$$\int_{\Gamma_2} P^i_{(0)} = 0,$$

for each effective 2-form $P^i_{(0)}$ of type *zero* on M. But, by § 45·1,

$$\int_{\Gamma_2} P^i_{(0)} = \nu^{-1}_{m-2}\int_{\Gamma_{2m-2}} P^i_{(0)} \times \omega_{m-2},$$

and it follows the necessary and sufficient condition that Γ_{2m-2} be algebraic is that the harmonic integrals of multiplicity $2m-2$ and type $m-2$ should have zero periods on Γ_{2m-2}. But, as in § 47·3, we show that any closed $(2m-2)$-form of type $m-2$ is homologous to a harmonic form of type $m-2$, and hence the necessary and sufficient condition that Γ_{2m-2} be algebraic is that the closed $(2m-2)$-forms of type $m-2$ have zero periods on it.

51·3. As an example of the use which can be made of the criterion that a cycle be algebraic, we refer briefly to the theory of correspondences between two irreducible curves C and D, of genera p and q respectively [13, 8]. We denote by x a general point of C, by $\gamma_1, \ldots, \gamma_{2p}$ a fundamental base for the 1-cycles of the Riemann surface of C, and we denote this Riemann surface by C. Similarly, $y, \delta_1, \ldots, \delta_{2q}, D$ refer to the 0-cycles, 1-cycles, and 2-cycle of the Riemann surface of D. Let $C \times D$ be the product of C by D, and denote its Riemannian manifold by the same symbol.

Suppose that we are given an algebraic correspondence T between C and D, in which to a point of C there correspond β points of D, and to a point of D there correspond α points of C. Let γ_i be transformed into $T(\gamma_i)$ of D, where

$$T(\gamma_i) \sim b^j_i \delta_j, \tag{6}$$

and similarly let δ_i be transformed by the reverse transformation into $T^{-1}(\delta_i)$ of C, where

$$T^{-1}(\delta_i) \sim a^j_i \gamma_j. \tag{7}$$

On the surface $C \times D$ the correspondence is "represented" by the curve which represents the point-pairs $x \times y$ related by the correspondence. This curve is an algebraic 2-cycle Γ, and it can be proved that

$$\Gamma \sim \beta C \times y + \alpha x \times D + \epsilon^{ij}\gamma_i \times \delta_j,$$

where
$$\boldsymbol{\epsilon} = \boldsymbol{\gamma}\mathbf{b} = -(\boldsymbol{\delta}\mathbf{a})';$$

here
$$\mathbf{b} = (b_i^j), \quad \mathbf{a} = (a_i^j),$$

and $\boldsymbol{\gamma}$, $\boldsymbol{\delta}$ are the transposes of the inverses of the intersection matrices of the 1-cycles on C, D.

Let
$$\int du_i \qquad (i = 1, \ldots, p),$$
$$\int dv_i \qquad (i = 1, \ldots, q)$$

be the Abelian integrals of the first kind attached to C, D, respectively, and let their period matrices be $\boldsymbol{\omega}$, $\boldsymbol{\nu}$. Then the algebraic double integrals of the first kind on $C \times D$ are the pq integrals

$$\int du_i \times dv_j.$$

Since Γ is algebraic,

$$\int_\Gamma du_i \times dv_j = 0 \qquad (i = 1, \ldots, p;\ j = 1, \ldots, q),$$

from which we obtain the matrix equation

$$\boldsymbol{\omega\epsilon\nu}' = \mathbf{0}.$$

This is, essentially, a classical result due to Hurwitz [6].

Conversely, suppose there exists a relation

$$\boldsymbol{\omega\epsilon\nu}' = \mathbf{0}$$

connecting $\boldsymbol{\omega}$, $\boldsymbol{\nu}$, where $\boldsymbol{\epsilon}$ is a matrix of integers. Then this condition implies that

$$\Gamma \sim \beta C \times y + \alpha x \times D + \epsilon^{ij}\gamma_i \times \delta_j$$

is algebraic, whatever values we give to α, β; and a simple geometrical argument is sufficient to show that α, β can be chosen so that the cycle represents an effective algebraic curve. This curve defines a correspondence between C and D in which the 1-cycles are transformed according to (6) and (7).

The correspondence is of valency zero if and only if

$$\boldsymbol{\epsilon} = \mathbf{0},$$

and the result shows that the number of independent "singular" correspondences between C and D is the number of independent matrices $\boldsymbol{\epsilon}$ for which

$$\boldsymbol{\omega\epsilon\nu}' = \mathbf{0}.$$

For a more detailed account of this theory, and for further developments, the reader is referred to a paper by Lefschetz [8]. A partial extension to correspondences between surfaces will be found in [5]. The reason that the extension is only partial is that we can only apply necessary conditions that a 4-cycle be algebraic.

52·1. **Some results for surfaces.** We conclude this chapter by referring to certain special results which we can deduce for algebraic surfaces. We first make a deduction from the fact that any algebraic integral of the first kind is harmonic, and therefore cannot have all its periods zero.

Let us consider an algebraic surface V_2 of order n, lying in a space of three dimensions and having ordinary singularities, whose equation in non-homogeneous coordinates is

$$f(x, y, z) = 0.$$

Severi [16] has defined a *semi-exact* integral on V_2 to be an integral of the form

$$\int R\,dx + S\,dy, \tag{8}$$

where R, S are rational functions on V_2, if the integral is such that on every curve of V_2 it defines an Abelian integral of the first kind. Clearly, such an integral is birationally invariant, and it can be expressed, in the neighbourhood of any point of V_2, in terms of local parameters u, v in the form

$$\int P\,du + Q\,dv,$$

where P, Q are analytic functions of u, v. Therefore

$$\int\left(\frac{dS}{dx} - \frac{dR}{dy}\right) dx\,dy,$$

which is equal, locally, to

$$\int\left(\frac{\partial Q}{\partial u} - \frac{\partial P}{\partial v}\right) du\,dv,$$

is finite everywhere on V_2, and, since it is the integral of a null form, has no periods. But the double integral is algebraic, and therefore, since it has no periods, we must have

$$\frac{dS}{dx} - \frac{dR}{dy} = 0.$$

In other words, every semi-exact integral is the integral of a closed form, that is, it is an algebraic simple integral of the first kind (usually called a Picard integral of the first kind) on V_2. Now let us consider the integral (8) on the curve $x = \text{constant}$. The function S must be a polynomial A, adjoint to $f(x, y, z)$, of degree $(n-3)$ in (y, z), over $\partial f/\partial z$, that is,

$$S = \frac{A}{f_z},$$

where we have written $f_z = \partial f/\partial z$. Similarly we may write

$$R = -\frac{B}{f_z},$$

where B is an adjoint polynomial, of degree $(n-3)$ in (x, z). By considering the behaviour of the integrals at infinity we see that A and B are of degree $(n-2)$ in (x, y, z).

Next, we consider what the integral becomes when we take

(y, z) as the independent variables. To do this we eliminate dx by means of the equation

$$f_x dx + f_y dy + f_z dz = 0,$$

and obtain
$$\int R\,dx + S\,dy = \int \frac{B\,dz - C\,dy}{f_x},$$

where
$$C = -\frac{Af_x + Bf_y}{f_z}.$$

Exactly as before, we prove that C can be taken to be equal, on V_2, to a polynomial of degree $(n-2)$ in (x, y, z) and of degree $(n-3)$ in (x, y) adjoint to $f(x, y, z)$, and we have a relation

$$Af_x + Bf_y + Cf_z = Df, \tag{9}$$

where D is a polynomial of degree $(n-3)$. The integral can also be written as

$$\int \frac{C\,dx - A\,dz}{f_y}.$$

The integrability condition is

$$\frac{\partial}{\partial x}\left(\frac{A}{f_z}\right) - \frac{f_x}{f_z}\frac{\partial}{\partial z}\left(\frac{A}{f_z}\right) + \frac{\partial}{\partial y}\left(\frac{B}{f_z}\right) - \frac{f_y}{f_z}\frac{\partial}{\partial z}\left(\frac{B}{f_z}\right) = 0,$$

and this reduces to
$$\frac{\partial A}{\partial x} + \frac{\partial B}{\partial y} + \frac{\partial C}{\partial z} = D \tag{10}$$

on V_2. Since each side of this equation is a polynomial of degree $(n-3)$, it follows that (10) is an identity.

Severi [16] has further shown that if there exist polynomials A, B, C, D, of the given degrees, A, B, C being adjoint to f, which satisfy (9), we necessarily have

$$A = x\phi + \alpha, \quad B = y\phi + \beta, \quad C = z\phi + \gamma, \quad D = n\phi + \delta,$$

where ϕ is a homogeneous polynomial of degree $(n-3)$, and α, β, γ are of degree $(n-3)$, while δ is of degree $(n-4)$. It now follows easily (Picard and Simart [10]) that equations (9) and (10) are necessary and sufficient to define a Picard integral of the first kind on V_2. Equation (9) is a sufficient condition in order that

$$\int \frac{B\,dz - C\,dy}{f_x}$$

should be a semi-exact integral. But we have just seen that a semi-exact integral is necessarily closed. Therefore (10) is satisfied, and equation (9) is sufficient to define a Picard integral of the first kind on the surface V_2.

Severi goes on to make an interesting deduction. The equation

$$A = 0$$

defines a surface of order $(n-2)$, adjoint to V_2, with the following properties. It passes through the points of V_2 which satisfy

$$f_y = 0, \quad f_z = 0,$$

that is, through the Jacobian set of the pencil of curves cut by the planes $x =$ constant. Moreover, it passes through the base points of this pencil. Severi shows that if A is any polynomial of degree $(n-2)$ which defines a surface

$$A = 0$$

satisfying these conditions, there exist unique polynomials B, C, D such that A, B, C, D satisfy (9). Hence the number of independent Picard integrals of the first kind is the number of linearly independent polynomials A. Stated geometrically, this result becomes:

Let $|E|$ *be a proper system of curves on* V_2, *and let* $|E_1|$ *be a pencil belonging to* $|E|$, *and* $|E'|$ *the system adjoint to* $|E|$. *The number of independent Picard integrals of the first kind on* V_2 *is equal to the number of curves of the complete system* $|E+E'|$ *which pass through* (a) *the Jacobian set of* $|E_1|$, (b) *the base points of* $|E_1|$.

Finally, Severi shows that the curves of $|E+E'|$ which pass through the set (a) all pass through the set (b).

It is a well-known theorem ([15] and [7]) that the number of independent Picard integrals of the first kind is $p_g - p_a$, where p_g is the geometric genus of V_2, and p_a its arithmetic genus.

52·2. We now consider a second application of the theory of harmonic integrals to algebraic surfaces. Let ρ be the maximum number of linearly independent 2-cycles of the surface V_2 which are algebraic, and let

$$\Gamma_2^0 = M(V_1), \quad \Gamma_2^1, \ldots, \Gamma_2^{\rho-1}$$

be a base for them, where Γ_2^i $(i \geq 1)$ are effective cycles. The intersection matrix of these cycles is non-singular (Severi [14]), and therefore, as in § 50, we can choose $R_2 - \rho$ further effective cycles so that $\Gamma_2^1, \ldots, \Gamma_2^{R_2-1}$ form a base for the effective 2-cycles of M, and

$$(\Gamma_2^i \,.\, \Gamma_2^j) = 0 \qquad (i < \rho;\ j \geq \rho).$$

A birational transformation affects only the algebraic cycles, hence the cycles $\Gamma_2^\rho, \ldots, \Gamma_2^{R_2-1}$ are absolutely invariant. They are called the *transcendental* cycles of M. We write the intersection matrix for the effective cycles as

$$\mathbf{a} = \begin{pmatrix} \mathbf{a}_1 & 0 \\ 0 & \mathbf{a}_2 \end{pmatrix},$$

where $\mathbf{a}_1$ is a symmetric matrix of $\rho - 1$ rows and columns, and $\mathbf{a}_2$ is a symmetric matrix of $R_2 - \rho$ rows and columns. Both are non-singular. Now let us write the effective integrals as a complete set, viz.

(i) the algebraic integrals $\quad \int P_i \qquad (i = 1, \ldots, p_g),$

where $p_g = \rho_2^0$ is the geometric genus of V_2;

(ii) $$\int Q_i \qquad (i = 1, \ldots, R_2 - 1 - 2p_g),$$

where $$Q_i = \bar{Q}_i;$$

(iii) $$\int R_i \qquad (i = 1, \ldots, p_g),$$

where $$R_i = \bar{P}_i,$$

and let the period matrices of the three sets be $\mathbf{\Omega}$, $\mathbf{\Lambda}$, $\bar{\mathbf{\Omega}}$.

The matrix $\mathbf{\Lambda}$ satisfies the equations

$$\mathbf{\Omega}(\mathbf{a}^{-1})'\mathbf{\Lambda}' = 0,$$

$$\bar{\mathbf{\Omega}}(\mathbf{a}^{-1})\mathbf{\Lambda}' = 0,$$

but, since $\mathbf{\Lambda}$ and $\mathbf{a}$ are real, we need only consider the first set,

$$\mathbf{\Omega}(\mathbf{a}^{-1})'\mathbf{\Lambda}' = 0.$$

Any real set of solutions of

$$\mathbf{\Omega}(\mathbf{a}^{-1})'\mathbf{x} = 0$$

is a set of periods of a linear combination of the integrals (ii).

Since the first $\rho-1$ cycles are algebraic, we have, by Lefschetz's theorem (§ 51·2),

$$\mathbf{\Omega} = (0, \boldsymbol{\omega}),$$

where $\boldsymbol{\omega}$ has $R_2-\rho$ columns. The equations to determine $\mathbf{\Lambda}$ are therefore

$$(0, \boldsymbol{\omega}[\mathbf{a}_2^{-1}]')\mathbf{\Lambda} = 0.$$

Therefore, by rearranging the integrals (ii) we can take $\mathbf{\Lambda}$ to be of the form

$$\mathbf{\Lambda} = \begin{pmatrix} \mathbf{I}_{\rho-1} & \cdot \\ \cdot & \mathbf{\Lambda}_1 \end{pmatrix},$$

where $\mathbf{I}_{\rho-1}$ is the unit matrix of $\rho-1$ rows and columns. Hence the effective integrals form a defective set, and can be resolved into the two complementary complete sets:

(*a*) $\int Q_i \quad (i=1, \ldots, \rho-1);$

(*b*) (i) $\int P_i \quad (i=1, \ldots, p_g),$

(ii) $\int Q_i \quad (i=\rho, \ldots, R_2-1-2p_g),$

(iii) $\int R_i \quad (i=1, \ldots, p_g).$

Now apply to these two sets the theorem of § 47·1. We obtain the results

(*a*) $\mathbf{a}_1$ is a positive definite matrix;

(*b*)
$$\begin{pmatrix} \mathbf{\Omega} \\ \mathbf{\Lambda}_1 \\ \bar{\mathbf{\Omega}} \end{pmatrix} (\mathbf{a}_2^{-1})' (\bar{\mathbf{\Omega}}' \mathbf{\Lambda}_1' \mathbf{\Omega}') = \begin{pmatrix} -\mathbf{b}_1 & 0 & 0 \\ 0 & \mathbf{b}_2 & 0 \\ 0 & 0 & -\mathbf{b}_3 \end{pmatrix},$$

where
$$\begin{pmatrix} \mathbf{b}_1 & 0 & 0 \\ 0 & \mathbf{b}_2 & 0 \\ 0 & 0 & \mathbf{b}_3 \end{pmatrix}$$

is a positive definite Hermitian matrix. Hence $\mathbf{a}_2$ is a matrix whose signature is $R_2 - \rho - 2p_g$. Thus $\mathbf{a}_1$, $\mathbf{a}_2$ are symmetric matrices whose signatures are $\rho - 1$, $R_2 - \rho - 2p_g$, respectively.

In interpreting these results, the reader should remember that if V_2 is of order n,

$$(\Gamma_2^0 \,.\, \Gamma_2^0) = -n.$$

In other words, our conventions of orientation are such that a positive geometrical intersection corresponds to a negative topological intersection. It is therefore convenient in connection with the present results to change the orientation of M so that the parameters (x_1, x_3, x_2, x_4) are concordant with the orientation. The first theorem becomes:

I. *The signature of the intersection matrix of the transcendental* 2-*cycles of* M *is* $2p_g$.

For the algebraic cycles, we now consider the intersection matrix of the ρ cycles $\Gamma_2^0, \ldots, \Gamma_2^{\rho-1}$. Then we have

II. *The intersection matrix of the curves of a base on an algebraic cycle has signature* 1. Geometrical proofs of this last theorem have been given by Segre [12] and Bronowski [1].

REFERENCES

1. J. Bronowski. *Journal of the London Mathematical Society*, 13 (1938), 86.
2. G. Castelnuovo and F. Enriques. *Annales de l'École Normale Supérieure* (3), 23 (1906), 339.
3. L. E. Dickson. *Modern Algebraic Theories* (Chicago), 1926, p. 71.
4. W. V. D. Hodge. *Journal of the London Mathematical Society*, 12 (1937), 280.
5. W. V. D. Hodge. *Proceedings of the London Mathematical Society* (2), 44 (1938), 226.
6. A. Hurwitz. *Mathematische Annalen*, 28 (1887), 561.
7. S. Lefschetz. *L'Analysis Situs et la Géométrie Algébrique* (Paris), 1924.
8. S. Lefschetz. *Annals of Mathematics*, 28 (1927), 342.
9. G. Mannoury. *Nieuw Archief voor Wiskunde* (2), 4 (1898), 112.
10. E. Picard and G. Simart. *Théorie des Fonctions Algébriques de deux Variables*, vol. 1 (Paris), 1897.
11. H. Poincaré. *American Journal of Mathematics*, 8 (1886), 289.
12. B. Segre. *Annali di Matematica* (4), 16 (1937), 157.
13. F. Severi. *Memorie della Reale Accademia di Torino* (2), 14 (1904), 1.
14. F. Severi. *Mathematische Annalen*, 62 (1906), 194.
15. F. Severi. *Atti della Reale Accademia nazionale Lincei* (5), 30 (1921).
16. F. Severi. *Atti della Reale Accademia nazionale Lincei* (6), 7 (1928).
17. O. Zariski. *Algebraic Surfaces* (Berlin), 1935.

Chapter V

APPLICATIONS TO THE THEORY OF CONTINUOUS GROUPS

In this chapter we consider the application of the theory of harmonic integrals to spaces which possess continuous groups of transformations into themselves. It will be shown that when a space possesses a continuous group of transformations into itself, and the group fulfils certain general conditions, a Riemannian metric can be defined in the space by means of this group, and the harmonic integrals associated with this metric are invariants of the group. It will be shown that these invariants are, in fact, the same as certain invariants which have been discussed by Cartan [2]. The main purpose of this chapter is to show how harmonic integrals provide a convenient method of discussing the invariants of Cartan.

53. **Continuous groups.** For the purposes of this chapter, some knowledge of the elements of the theory of continuous groups must be assumed. The results which the reader will require to know can be found in any standard treatise on continuous groups. For his convenience we shall make our references for results in Lie's theory of groups to the work of Eisenhart [3], and for results taken from the topological theory of groups the reader may consult the monograph of Cartan [1]. But in order to explain our approach to the subject, and to fix our notation, we shall begin this chapter with a summary of the theory which we propose to make the basis of our later investigations. It should be noted that in some respects we impose certain limitations on the spaces and groups considered which are unnecessary in the general theory of continuous groups; the reasons for these restrictions will, however, be apparent later.

53·1. Let V be an orientable manifold of class u, of n dimensions. Since the only transformations of coordinates on V which we have to consider are analytic transformations, the class number u is of no importance, and we may, if we wish, suppose that V is an analytic manifold. The condition that V should be an orientable manifold is essential for the theory which will come later.

A transformation T of the space V is a correspondence which relates each point P of V to another point P' of V, in such a way that the correspondence between P and P' is one to one, without exception. As P describes the whole manifold V, P' also describes the whole manifold, and, given any point Q of V, there is just one point Q' of V such that, when P' is at Q, P is at Q'. The transformation which takes Q into Q' is called the *inverse* of T and is denoted by T^{-1}. The transformation which takes each point of V into itself is called the *identity* transformation.

Instead of considering a single transformation T of V, we now consider a set of transformations of V, which are in one to one correspondence with the points of a space M. For the purposes of this chapter, we confine ourselves to the case in which M is a manifold of r dimensions. It will appear later that the manifolds M which we have to consider are of dimension *three* at least. If A is any point of M, we denote the corresponding transformation by T_A. The set of transformations is said to form an r-parameter continuous group if the following properties are satisfied:

(i) if T_A is any transformation of the set, the inverse transformation T_A^{-1} also belongs to the set;

(ii) if T_A and T_B are any two transformations of the set, and T_A takes the point P of V into the point P', and T_B takes the point P' into P'', there is a transformation of the set which takes P into P'', for all points P of V. This transformation is called the *product* of the transformation T_A by T_B, and is denoted by $T_B T_A$;

(iii) the rule for forming the products of transformations obeys the associative law, that is, if

$$T_B T_A = T_X \quad \text{and} \quad T_C T_B = T_Y,$$

then
$$T_C T_X = T_Y T_A.$$

Throughout this chapter, we assume that the sets of transformations which we consider satisfy these conditions. We further assume that the following conditions are satisfied. Let $(s_1, \ldots, s_r)$ be a local coordinate system on M valid in some neighbourhood, and let A be any point of this neighbourhood. The transformations T_A transform the points of a neighbourhood N of V into points of a neighbourhood $\overline{N}$ of V. If $(x_1, \ldots, x_n)$ is a coordinate system valid in N, and $(\bar{x}_1, \ldots, \bar{x}_n)$ is a coordinate system valid in $\overline{N}$, the transformation T_A, where A has coordinates $(s_1, \ldots, s_r)$, transforms the point P, whose coordinates are $(x_1, \ldots, x_n)$, into the point P' whose coordinates $(\bar{x}'_1, \ldots, \bar{x}'_n)$ are given by

$$\bar{x}'_i = f_i(x_1, \ldots, x_n; s_1, \ldots, s_r) \qquad (i = 1, \ldots, n), \quad (1)$$

the functions being real analytic functions of $(x_1, \ldots, x_n)$ and of $(s_1, \ldots, s_r)$, and the determinant $\left|\frac{\partial f_i}{\partial x_j}\right|$ being different from zero at any point of N, for all positions of A.

Again, let $(s_1, \ldots, s_r)$, $(\bar{s}_1, \ldots, \bar{s}_r)$ be local coordinate systems on M, valid in neighbourhoods M_1 and M_2. If A is any point of M_1, and B any point of M_2, we assume that as A and B vary in their neighbourhoods the product transformation $T_C = T_B T_A$ defines a set of points C on M which lie in a neighbourhood M_3. If $(s'_1, \ldots, s'_r)$ is a local coordinate system valid in M_3, C is determined by the equations

$$s'_i = \phi_i(s_1, \ldots, s_r; \bar{s}_1, \ldots, \bar{s}_r) \qquad (i = 1, \ldots, r), \quad (2)$$

where the functions are analytic in $(s_1, \ldots, s_r)$ and in $(\bar{s}_1, \ldots, \bar{s}_r)$.

Finally, we assume that the group of transformations on V is *transitive*, that is, that there exists a transformation of the group which transforms any assigned point P of V into any assigned point P' of V. This implies that $r \geqslant n$.

The conditions which we have imposed imply that the groups to be considered are of the kind known in group theory as transitive, closed, Lie groups.

53·2. By using the properties of groups which we have stated above, it is shown in the standard works ([3], p. 18) that if we eliminate $(x_1, \ldots, x_n)$ from equations (1) and the equations

$$\frac{\partial \bar{x}'_i}{\partial s_\alpha} = \frac{\partial}{\partial s_\alpha} f_i(x, s),$$

we obtain
$$\frac{\partial \bar{x}'_i}{\partial s_\alpha} = \xi^i_a(\bar{x}')\, A^a_\alpha(s), \tag{3}$$

where, for each a, $\xi^i_a(\bar{x}')$ is a contravariant vector at the point $(\bar{x}'_1, \ldots, \bar{x}'_n)$ on V which is independent of $(s_1, \ldots, s_r)$, the components of the vector being given in the coordinate system $(\bar{x}_1, \ldots, \bar{x}_n)$; and where $A^a_\alpha(s)$ is a covariant vector at the point $(s_1, \ldots, s_r)$ on M, which is independent of $(\bar{x}'_1, \ldots, \bar{x}'_n)$. The summation is from $a = 1$ to $a = r$. The vectors ξ^i_a $(a = 1, \ldots, r)$ are linearly independent, in the sense that there exist no constants $c^1, \ldots, c^r$, different from zero, such that

$$c^a \xi^i_a = 0.$$

Similarly the vectors A^a_α are linearly independent and the determinant $| A^a_\alpha |$ is never zero. Hence we can find a set of r contravariant vectors A^α_a $(a = 1, \ldots, r)$ on M such that

$$A^\alpha_a(s)\, A^b_\alpha(s) = \delta^b_a.$$

The vectors ξ^i_a are determined by the transformations of the group, as a set. But we can replace them by linear combinations (with constant coefficients) of themselves. Let

$$\eta^i_a = u^b_a \xi^i_b \qquad (a = 1, \ldots, r),$$

where u^b_a $(a, b = 1, \ldots, r)$ are real constants, and $| u^b_a | \neq 0$. We can find r^2 real numbers $v^a_b (a, b = 1, \ldots, r)$ such that

$$u^a_b v^b_c = \delta^a_c.$$

If we write
$$B^a_\alpha = v^a_b A^b_\alpha,$$
we have
$$\eta^i_a B^a_\alpha = u^b_a \xi^i_b v^a_c A^c_\alpha$$
$$= \xi^i_a A^a_\alpha,$$
and hence
$$\frac{\partial \bar{x}'_i}{\partial s_\alpha} = \eta^i_a(\bar{x}')\, B^a_\alpha(s).$$

The set of vectors ξ^i_a $(a=1, \ldots, r)$ may therefore be regarded as the components of a covariant vector in a space of r dimensions, which we shall call the vector space associated with the group. In this vector space the allowable transformations are those in which the coefficients in the equations of transformation of a vector are constants, and have a non-singular determinant. The r vectors A^a_α $(a=1, \ldots, r)$ on M then determine a contravariant vector in this space, and the vectors A^α_a determine a covariant vector.

The equations (3) are called the fundamental equations of the group, and the vectors ξ^i_a are called the fundamental covariant vectors on V. When the coordinate system in the associated vector space is fixed, there is a vector ξ^i_a defined at each point of V, uniquely for each a. Similarly A^a_α, A^α_a are respectively the fundamental covariant and contravariant vectors on M.

53·3. The fundamental equations of the group are differential equations which determine $(\bar{x}'_1, \ldots, \bar{x}'_n)$ as functions of $(s_1, \ldots, s_r)$, when the initial values are properly chosen. They are completely integrable. If we apply the conditions of integrability
$$\frac{\partial}{\partial s_\alpha}[\xi^i_a(\bar{x}')\, A^a_\beta(s)] = \frac{\partial}{\partial s_\beta}[\xi^i_a(\bar{x}')\, A^a_\alpha(s)],$$
we obtain the equations
$$\xi^j_a \frac{\partial \xi^i_b}{\partial \bar{x}'_j} - \xi^j_b \frac{\partial \xi^i_a}{\partial \bar{x}'_j} = C^c_{ab}\, \xi^i_c, \tag{4}$$
where
$$C^c_{ab} = A^\alpha_a A^\beta_b \left(\frac{\partial A^c_\alpha}{\partial s_\beta} - \frac{\partial A^c_\beta}{\partial s_\alpha}\right).$$

This last equation can also be written as

$$A^\beta_a \frac{\partial A^\alpha_b}{\partial s_\beta} - A^\beta_b \frac{\partial A^\alpha_a}{\partial s_\beta} = C^c_{ab} A^\alpha_c. \tag{5}$$

Since the vectors ξ^j_a do not depend on $(s_1, \ldots, s_r)$, and the vectors A^α_a, A^a_α do not depend on $(\bar{x}'_1, \ldots, \bar{x}'_n)$, it follows that C^c_{ab} is a constant. Clearly, we have

$$C^c_{ab} = -C^c_{ba},$$

and, since

$$0 = \sum_{a,b,c} \left\{ \xi^j_a \frac{\partial}{\partial \bar{x}'_j}\left[\xi^k_b \frac{\partial \xi^i_c}{\partial \bar{x}'_k} - \xi^k_c \frac{\partial \xi^i_b}{\partial \bar{x}'_k}\right] - \left[\xi^k_b \frac{\partial \xi^j_c}{\partial \bar{x}'_k} - \xi^k_c \frac{\partial \xi^j_b}{\partial \bar{x}'_k}\right] \frac{\partial \xi^i_a}{\partial \bar{x}'_j} \right\}$$

$$= [C^d_{bc} C^e_{ad} + C^d_{ca} C^e_{bd} + C^d_{ab} C^e_{cd}] \xi^i_e,$$

we have

$$C^d_{bc} C^e_{ad} + C^d_{ca} C^e_{bd} + C^d_{ab} C^e_{cd} = 0. \tag{6}$$

The numbers C^c_{ab} $(a, b, c = 1, \ldots, r)$ are called the *constants of structure* of the group. If we replace the vectors ξ^i_a by vectors $\eta^i_a = u^b_a \xi^i_b$, etc., we find that C^c_{ab} is replaced by

$$D^c_{ab} = u^p_a u^q_b v^c_r C^r_{pq}.$$

Thus C^c_{ab} is a tensor of the associated vector space.

Another tensor of the vector space is given by

$$g_{ab} = C^e_{ad} C^d_{eb}.$$

This is a symmetric tensor. It is shown in works on continuous groups ([3], p. 174) that for groups of the kind known as semi-simple, the determinant $|g_{ab}|$ is not zero. Semi-simple groups are usually defined in terms of properties of the invariant sub-groups of a group, and it is then shown that the necessary and sufficient condition that a group be semi-simple is that $|g_{ab}| \neq 0$. For our purposes, however, it is sufficient to define a semi-simple group as a group for which $|g_{ab}| \neq 0$, but it is to be noted that this condition defines a class of groups which is important in other connections. Cartan ([1], p. 10) has shown that for a semi-simple group which is closed the matrix (g_{ab}) is positive definite. *In this chapter we are only concerned with closed semi-simple groups.*

We take the tensor g_{ab} as the metrical tensor in the vector space. There exists a contravariant tensor g^{ab} such that

$$g^{ab}g_{bc} = \delta^a_c,$$

and the tensors g^{ab}, g_{bc} can be used to raise and lower indices. We shall write

$$C_{abc} = g_{ad}\, C^d_{bc}.$$

Now
$$\begin{aligned} C_{abc} &= C^p_{aq}\, C^q_{pd}\, C^d_{bc} \\ &= C^p_{aq}[C^q_{bd}\, C^d_{pc} + C^q_{cd}\, C^d_{bp}] \\ &= C^d_{pc}[C^p_{aq}\, C^q_{bd} - C^p_{bq}\, C^q_{ad}] \\ &= -C_{bac}. \end{aligned}$$

Hence C_{abc} is a skew-symmetric tensor of the vector space. We shall find it convenient in raising and lowering indices in tensors such as C^c_{ab} to observe the convention that when we raise an index which is on the left at the bottom we place it on the right at the top. Thus we write

$$g^{ad}C^c_{db} = C^{ca}_b.$$

With this convention we have, for instance,

$$g^{ad}C^c_{bd} = -g^{ad}C^c_{db} = -C^{ca}_b,$$

and it follows easily that

$$C^{ac}_b = -C^{ca}_b.$$

53·4. Let us now consider the manifold M, which we call the group manifold. Let A and B be any two points of M, and let $T_B T_A = T_C$. This relation defines a transformation of M which takes the point A into the point C. For each point B of M there exists such a transformation of M, and it can be verified that these transformations form a transitive group, and, since there is one transformation of this group corresponding to each point B of M, M is also the group manifold of this new group. We call this group of transformations of M the *left-hand translation group*.

The equations of the transformations of the left-hand translation group are the equations

$$s'_i = \phi_i(s_1, \ldots, s_r; \bar{s}_1, \ldots, \bar{s}_r) \qquad (i = 1, \ldots, r), \quad (2)$$

where $(s_1, \ldots, s_r)$, $(\bar{s}_1, \ldots, \bar{s}_r)$ and $(s'_1, \ldots, s'_r)$ are the coordinates of A, B, C, respectively. We can form the fundamental equations of the group, and it is shown ([3], p. 23) that

$$\frac{\partial s'_\alpha}{\partial \bar{s}_\beta} = A^\alpha_a(s')\, A^a_\beta(\bar{s}), \quad (7)$$

where A^α_a, A^a_α are the vectors defined in § 53·2. Forming the conditions of integrability of these equations, we obtain the equations

$$A^\beta_a \frac{\partial A^\alpha_b}{\partial s'_\beta} - A^\beta_b \frac{\partial A^\alpha_a}{\partial s'_\beta} = C^c_{ab} A^\alpha_c, \quad (5)$$

which we have obtained above. Thus the left-hand translation group has the same constants of structure as the group of transformations of V.

Equations (7) can be used to show that the manifold M is orientable. To see this, let us fix some point of M, say the point O which represents the identity transformation, and let N be a neighbourhood of O, and $(s_1, \ldots, s_r)$ a coordinate system in N. Consider the set of neighbourhoods N_T of M, where the points of N_T represent the transformations $T_A T$ as A describes N. There is one such neighbourhood for each transformation T of the group, and the set of neighbourhoods covers M. By the theorem of Heine-Borel, there exists a finite number of transformations $T_1, \ldots, T_\rho$ such that the neighbourhoods $N_i = N_{T_i}$ $(i = 1, \ldots, \rho)$ cover M. We fix a coordinate system in N_i by the condition that a point P of N_i has the same coordinates as the point of N which represents the transformation $T_P T_i^{-1}$. If we can show that any two of these coordinate systems which are valid in a neighbourhood are like in this neighbourhood, it will follow that M is orientable.

Let P be any point common to N_i and N_j, and let the transformations $T_P T_i^{-1}$, $T_P T_j^{-1}$ be represented by the points A_i and

A_j, which lie in N. If A_i has the coordinates $(s_1^i, \ldots, s_r^i)$ and if A_j has the coordinates $(s_1^j, \ldots, s_r^j)$, the law of transformation expressing the coordinates in N_j in terms of the coordinates in N_i is the particular case of (2),

$$s_\alpha^j = \phi_\alpha(\sigma, s^i),$$

where σ represents the transformation $T_i T_j^{-1}$. Hence, by (7), we have

$$\frac{\partial s_\alpha^j}{\partial s_\beta^i} = A_a^\alpha(s^j)\, A_\beta^a(s^i),$$

and therefore
$$\left|\frac{\partial s_\alpha^j}{\partial s_\beta^i}\right| = |\, A_a^\alpha \,|_{A_j} \Big/ |\, A_a^\alpha \,|_{A_i}.$$

Now the determinant $|\, A_a^\alpha \,|$ has the same sign at all points of N, and hence

$$\left|\frac{\partial s_\alpha^j}{\partial s_\beta^i}\right| > 0,$$

and it follows that the coordinate systems in N_i, N_j are like in a neighbourhood of P. Thus M is orientable.

53·5. The equation
$$T_B T_A = T_C$$

serves to define another group of transformations on M, which we call the *right-hand translation group*. This group is composed of the transformations which take the point B into the point C, one transformation corresponding to each point A of M. The fundamental equations of this group are

$$\frac{\partial s'_\alpha}{\partial s_\beta} = \bar{A}_a^\alpha(s')\, \bar{A}_\beta^a(s),$$

where the vectors $\bar{A}_a^\alpha$, $\bar{A}_\alpha^a$ can be determined without difficulty ([3], p. 28). We choose a base for the set of vectors $\bar{A}_a^\alpha$ so that, at the point O which represents the identity transformation,

$$\bar{A}_a^\alpha = A_a^\alpha.$$

Then the vectors $\bar{A}^{\alpha}_{a}$, $\bar{A}^{a}_{\alpha}$ have the properties:

(i) $$\bar{A}^{\alpha}_{a}(s)\,\bar{A}^{b}_{\alpha}(s) = \delta^{b}_{a};$$

and (ii) $$\bar{A}^{\alpha}_{a}\frac{\partial \bar{A}^{a}_{\gamma}}{\partial s_{\beta}} = A^{\alpha}_{a}\frac{\partial A^{a}_{\beta}}{\partial s_{\gamma}};$$

and indeed the vectors $\bar{A}^{\alpha}_{a}$, $\bar{A}^{a}_{\alpha}$ can be defined by these properties.

The conditions of integrability of the fundamental equations of the right-hand translation group are found to give the equations

$$\bar{A}^{\beta}_{a}\frac{\partial \bar{A}^{\alpha}_{b}}{\partial s_{\beta}} - \bar{A}^{\beta}_{b}\frac{\partial \bar{A}^{\alpha}_{a}}{\partial s_{\beta}} = \bar{C}^{c}_{ab}\,\bar{A}^{\alpha}_{c},$$

where $$\bar{C}^{c}_{ab} + C^{c}_{ab} = 0.$$

53·6. The results which have been sketched in the preceding paragraphs, and which will be found proved in detail in any textbook on the theory of continuous groups, include everything required for use in this chapter. We are now ready to begin the particular study of this chapter, which is the behaviour of tensors on a manifold possessing a transitive closed semi-simple group of transformations, under transformations of the group. In particular we are interested in the skew-symmetric covariant tensors which are invariant under the transformations of the group. These define invariant integrals of the group. It is shown that the group defines a Riemannian metric on the manifold, and that the harmonic integrals associated with this metric are amongst the invariant integrals, and in the case of the translation groups on a group manifold we are able to deduce in this way certain topological properties of the manifold.

We shall be concerned with tensors on V, tensors on M, and tensors in the associated vector space. However, we never have to consider V and M simultaneously, though we have to consider each in turn in conjunction with the vector space. Thus, we are at any stage only concerned with two sets of tensors, and it is convenient to make certain conventions with regard to notation. When we are dealing with the manifold V and the vector space, we shall use Latin letters for the indices of tensors in V, Greek letters for the indices of tensors in the vector space. Thus, for ξ^i_a, C^c_{ab}, we shall write ξ^i_α, $C^\gamma_{\alpha\beta}$.

When we consider the group manifold M, we are only concerned with the translation groups, and for these groups V and M coincide. The fundamental contravariant vectors for the left-hand translation group are the vectors A^α_a, and, regarding the manifold as the manifold V, we replace these by ξ^i_a. Then we adopt the conventional use of Latin and Greek suffixes explained above, and write the vectors as ξ^i_α. We also have to consider the right-hand translation group. We shall denote the fundamental contravariant vectors of this group by η^i_α, these vectors being the vectors previously denoted by $\bar{A}^\alpha_a$.

54·1. Geometry of the transformation space. In this paragraph we are only concerned with the variety V. The infinitesimal transformations of the group are given by $x \to x'$, where

$$x'_i = x_i + \epsilon^\alpha \xi^i_\alpha,$$

ϵ^α being an infinitesimal number. Since the group is locally transitive, there is an infinitesimal transformation which takes a point x into any point of its neighbourhood, and for this it is necessary and sufficient that the matrix (ξ^i_α) should be of rank n. If we define g^{ij} by

$$g^{ij} = \xi^i_\alpha \xi^j_\beta g^{\alpha\beta},$$

the matrix (g^{ij}) is positive definite, and there therefore exists a positive definite matrix (g_{ij}) such that

$$g_{ij} g^{jk} = \delta^k_i.$$

The Riemannian metric on V is now defined by the quadratic differential form $g_{ij} dx^i dx^j$. It will be observed that the metric is determined completely by the structure of the continuous group. We use g_{ij} and g^{ij} to lower and raise Latin indices, and $g_{\alpha\beta}$ and $g^{\alpha\beta}$ to lower and raise Greek indices.

Consider $$\xi^\alpha_i = g^{\alpha\beta} g_{ij} \xi^j_\beta.$$

We have $$\xi^i_\alpha \xi^\alpha_j = \xi^i_\alpha g^{\alpha\beta} g_{jk} \xi^k_\beta = g^{ik} g_{jk} = \delta^i_j.$$

If $r = n$, we have $$\xi^i_\alpha \xi^\beta_i = \delta^\beta_\alpha,$$

but if $r > n$, $$h^\alpha_\beta = \xi^\alpha_i \xi^i_\beta \neq \delta^\alpha_\beta,$$

since (h^α_β) is only of rank n. The tensor h^α_β of the associated vector space is of great importance in discussing the geometry of V. It has the properties:

(i) $$h^\alpha_\beta \xi^i_\alpha = \xi^\alpha_j \xi^j_\beta \xi^i_\alpha = \delta^i_j \xi^j_\beta = \xi^i_\beta;$$

(ii) $$h^\alpha_\beta \xi^\beta_i = \xi^\alpha_i;$$

(iii) $$h^\alpha_\beta h^\beta_\gamma = h^\alpha_\beta \xi^\beta_i \xi^i_\gamma = \xi^\alpha_i \xi^i_\gamma = h^\alpha_\gamma.$$

Further, if $$h^{\alpha\beta} = g^{\alpha\gamma} h^\beta_\gamma,$$

then $$h^{\alpha\beta} = g^{\alpha\gamma} \xi^\beta_i \xi^i_\gamma = \xi^\beta_i g^{ij} \xi^\alpha_j = h^{\beta\alpha}.$$

Similarly, $$h_{\alpha\beta} = g_{\alpha\gamma} h^\gamma_\beta = h_{\beta\alpha},$$

and $$h^{\alpha\beta} h_{\beta\gamma} = h^\alpha_\gamma.$$

54·2. We now define quantities

$$L^i_{jk} = \xi^i_\alpha \frac{\partial \xi^\alpha_k}{\partial x_j} = -\xi^\alpha_k \frac{\partial \xi^i_\alpha}{\partial x_j}.$$

The law of transformation of these quantities corresponding to a change of coordinate system on V is easily found to be

$$\frac{\partial x_i}{\partial \bar{x}_a} \bar{L}^a_{jk} = \frac{\partial^2 x_i}{\partial \bar{x}_j \partial \bar{x}_k} + L^i_{bc} \frac{\partial x_b}{\partial \bar{x}_j} \frac{\partial x_c}{\partial \bar{x}_k}.$$

We may therefore take these quantities as the components of an affine connection on V. Since we shall consider more than one affine connection on V, we shall write the covariant derivative of a tensor $Q^{j_1\ldots j_q}_{i_1\ldots i_p}$ of weight W with respect to this connection as

$$Q^{j_1\ldots j_q}_{i_1\ldots i_p|k} = \frac{\partial}{\partial x_k} Q^{j_1\ldots j_q}_{i_1\ldots i_p} - \sum_{r=1}^{p} Q^{j_1\ldots\ldots\ldots\ldots\ldots j_q}_{i_1\ldots i_{r-1} a\, i_{r+1}\ldots i_p} L^a_{i\,k}$$

$$+ \sum_{r=1}^{q} Q^{j_1\ldots j_{r-1} a\, j_{r+1}\ldots j_q}_{i_1\ldots\ldots\ldots\ldots\ldots i_p} L^{j_r}_{ak} - W Q^{j_1\ldots j_q}_{i_1\ldots i_p} L^a_{ak}.$$

We shall also consider covariant differentiation with respect to the connection whose components are L^{*i}_{jk}, where

$$L^{*i}_{jk} - L^i_{kj} = 0,$$

and in this case we shall denote the covariant derivative of $Q^{j_1\ldots j_q}_{i_1\ldots i_p}$ by $Q^{j_1\ldots j_q}_{i_1\ldots i_p\|k}$.

Consider the fundamental relation satisfied by the vectors ξ^i_α,

$$\xi^j_\alpha \frac{\partial \xi^i_\beta}{\partial x_j} - \xi^j_\beta \frac{\partial \xi^i_\alpha}{\partial x_j} = C^\gamma_{\alpha\beta} \xi^i_\gamma. \tag{4}$$

If we multiply this by ξ^α_k, we obtain

$$\frac{\partial \xi^i_\beta}{\partial x_k} + \xi^j_\beta L^i_{jk} = C^\gamma_{\alpha\beta} \xi^i_\gamma \xi^\alpha_k,$$

that is,

$$\xi^i_{\beta|k} = C^\gamma_{\alpha\beta} \xi^i_\gamma \xi^\alpha_k. \tag{8}$$

Multiply again by ξ^β_l, and we obtain

$$L^i_{lk} - L^i_{kl} = C^\gamma_{\alpha\beta} \xi^i_\gamma \xi^\alpha_k \xi^\beta_l. \tag{9}$$

From (8) and (9), we have

$$\begin{aligned} \xi^i_{\beta\|k} &= C^\gamma_{\alpha\beta} \xi^i_\gamma \xi^\alpha_k - \xi^j_\beta (L^i_{jk} - L^i_{kj}) \\ &= C^\gamma_{\alpha\epsilon} \xi^i_\gamma \xi^\alpha_k (\delta^\epsilon_\beta - h^\epsilon_\beta). \end{aligned} \tag{10}$$

Expressions such as $C^\gamma_{\alpha\epsilon} \xi^i_\gamma \xi^\alpha_k$ occur so often in the sequel that it will be convenient to write them as $C^i_{k\epsilon}$, etc. Thus

$$C^i_{\alpha\beta} = \xi^i_\gamma C^\gamma_{\alpha\beta}, \quad C^\gamma_{i\beta} = C^\gamma_{\alpha\beta} \xi^\alpha_i,$$

and so on.

54·3. Let us consider the covariant derivatives of the metrical tensor.

$$\begin{aligned} g^{ij}{}_{|k} &= g^{\alpha\beta}\,\xi^i_{\alpha|k}\,\xi^j_\beta + g^{\alpha\beta}\xi^i_\alpha\,\xi^j_{\beta|k} \\ &= g^{\alpha\beta}C^i_{k\alpha}\,\xi^j_\beta + g^{\alpha\beta}C^j_{k\beta}\,\xi^i_\alpha \\ &= C^{ji}_k + C^{ij}_k = 0. \end{aligned}$$

Similarly,

$$\begin{aligned} g^{ij}{}_{\|k} &= g^{\alpha\beta}C^i_{k\epsilon}\,\xi^j_\beta(\delta^\epsilon_\alpha - h^\epsilon_\alpha) + g^{\alpha\beta}C^j_{k\epsilon}\,\xi^i_\alpha(\delta^\epsilon_\beta - h^\epsilon_\beta) \\ &= C^i_{k\epsilon}\,\xi^j_\beta(g^{\beta\epsilon} - h^{\beta\epsilon}) + C^j_{k\epsilon}\,\xi^i_\alpha(g^{\alpha\epsilon} - h^{\alpha\epsilon}) \\ &= 0, \end{aligned}$$

since $$\xi^j_\beta(g^{\beta\epsilon} - h^{\beta\epsilon}) = g^{jk}(\xi^\epsilon_k - \xi^\beta_k h^\epsilon_\beta) = 0.$$

If we write $$\Gamma^i_{jk} = \tfrac{1}{2}[L^i_{jk} + L^i_{kj}],$$

and if covariant differentiation with respect to the symmetric connection Γ^i_{jk} is written with the comma notation, we have, by adding the two results,

$$g^{ij}{}_{,k} = 0.$$

Hence $$g_{ij,k} = 0.$$

It follows that the components Γ^i_{jk} are the Christoffel symbols associated with the metric

$$g_{ij}\,dx^i\,dx^j.$$

From (8) and (10), we have

$$\xi^i_{\alpha,k} = C^i_{k\beta}(\delta^\beta_\alpha - \tfrac{1}{2}h^\beta_\alpha), \tag{11}$$

and, as a corollary,

$$\xi^i_{\alpha,i} = C^i_{i\beta}(\delta^\beta_\alpha - \tfrac{1}{2}h^\beta_\alpha) = h^\delta_\gamma\,C^\gamma_{\delta\beta}(\delta^\beta_\alpha - \tfrac{1}{2}h^\beta_\alpha) = 0.$$

Next,
$$\begin{aligned} \xi^\alpha_{i,k} &= g^{\alpha\beta}g_{ij}\,C^j_{k\gamma}(\delta^\gamma_\beta - \tfrac{1}{2}h^\gamma_\beta) \\ &= g_{ij}\,C^{\gamma j}_k(\delta^\alpha_\gamma - \tfrac{1}{2}h^\alpha_\gamma) \\ &= g_{\beta\epsilon}\,\xi^\epsilon_i\,C^{\gamma\beta}_k(\delta^\alpha_\gamma - \tfrac{1}{2}h^\alpha_\gamma) \\ &= C^\gamma_{ik}(\delta^\alpha_\gamma - \tfrac{1}{2}h^\alpha_\gamma). \end{aligned} \tag{12}$$

Similarly, we prove that

$$\frac{\partial}{\partial x_k} h^{\alpha}_{\beta} = C^{\alpha}_{\gamma k} h^{\gamma}_{\beta} - C^{\gamma}_{\beta k} h^{\alpha}_{\gamma}. \tag{13}$$

Finally, a useful formula is obtained by multiplying (4) by

$$(h^{\alpha}_{\lambda} - \delta^{\alpha}_{\lambda})(h^{\beta}_{\mu} - \delta^{\beta}_{\mu})\, \xi^{\nu}_{i}.$$

Since

$$(h^{\alpha}_{\lambda} - \delta^{\alpha}_{\lambda})\, \xi^{j}_{\alpha} = \xi^{j}_{\lambda} - \xi^{j}_{\lambda} = 0,$$

the left-hand side is zero, and we have

$$(h^{\alpha}_{\lambda} - \delta^{\alpha}_{\lambda})(h^{\beta}_{\mu} - \delta^{\beta}_{\mu})\, C^{\gamma}_{\alpha\beta} h^{\nu}_{\gamma} = 0. \tag{14}$$

55. **The transformation of tensors.** When a transformation of the group is applied to V, a tensor on V is transformed into another tensor. We are mainly interested in the conditions to be satisfied in order that a tensor be invariant, and for this it is sufficient to consider invariance under infinitesimal transformations, since a finite transformation can be built up from these. We therefore consider an infinitesimal transformation $x \to x'$, where

$$x'_i = x_i + \epsilon^{\alpha} \xi^{i}_{\alpha}, \tag{15}$$

where we can neglect squares and products of ϵ^{α}. To calculate the effect of this infinitesimal transformation on a tensor $Q^{j_1 \ldots j_q}_{i_1 \ldots i_p}$ of weight W, we proceed as follows. We first regard (15) as a change of coordinate system at the point x. Then, after the transformation (15) has been applied to the tensor, the components of the new tensor are given at x' in the original coordinate system by the values of the corresponding components of the original tensor at x in the new coordinate system. The difference between this and the value of the original tensor at the point x' gives the change produced in the given tensor at x' by the transformation. This change only differs from the change at x by quantities of the second order, and hence we have the expression for the change at x in the tensor produced by the transformation. We denote the given tensor at x by $Q^{j_1 \ldots j_q}_{i_1 \ldots i_p}(x)$, and the new tensor at the same point by $Q'^{j_1\ldots j_q}_{i_1\ldots i_p}(x)$. We then have (all in the original coordinate system):

$$Q'^{j_1\ldots j_q}_{i_1\ldots i_p}(x') = \left|\frac{\partial x_i}{\partial x'_j}\right|^W Q^{b_1\ldots b_q}_{a_1\ldots a_p}(x)\frac{\partial x_{a_1}}{\partial x'_{i_1}}\cdots\frac{\partial x_{a_p}}{\partial x'_{i_p}}\frac{\partial x'_{j_1}}{\partial x_{b_1}}\cdots\frac{\partial x'_{j_q}}{\partial x_{b_q}}$$

$$= \left[1 - W\epsilon^\alpha\frac{\partial \xi^k_\alpha}{\partial x_k}\right]\left[Q^{j_1\ldots j_q}_{i_1\ldots i_p}(x) + \epsilon^\alpha\sum_{r=1}^{q} Q^{j_1\ldots j_{r-1}a\,j_{r+1}\ldots j_q}_{i_1\ldots\ldots\ldots\ldots\ldots i_p}\frac{\partial \xi^{j_r}_\alpha}{\partial x_a} - \epsilon^\alpha\sum_{r=1}^{p} Q^{j_1\ldots\ldots\ldots\ldots\ldots j_q}_{i_1\ldots i_{r-1}a\,i_{r+1}\ldots i_p}\frac{\partial \xi^a_\alpha}{\partial x_{i_r}}\right]$$

$$= Q^{j_1\ldots j_q}_{i_1\ldots i_p}(x) - \epsilon^\alpha\left[\sum_{r=1}^{p} Q^{j_1\ldots\ldots\ldots\ldots\ldots j_q}_{i_1\ldots i_{r-1}a\,i_{r+1}\ldots i_p}\frac{\partial \xi^a_\alpha}{\partial x_{i_r}} - \sum_{r=1}^{q} Q^{j_1\ldots j_{r-1}a\,j_{r+1}\ldots j_q}_{i_1\ldots\ldots\ldots\ldots\ldots i_p}\frac{\partial \xi^{j_r}_\alpha}{\partial x_a} + W\frac{\partial \xi^k_\alpha}{\partial x_k}Q^{j_1\ldots j_q}_{i_1\ldots i_p}\right].$$

Hence $\quad Q'^{j_1\ldots j_q}_{i_1\ldots i_p}(x') - Q^{j_1\ldots j_q}_{i_1\ldots i_p}(x') = -\epsilon^\alpha \Delta_\alpha Q^{j_1\ldots j_q}_{i_1\ldots i_p},$

where

$$\Delta_\alpha Q^{j_1\ldots j_q}_{i_1\ldots i_p} = \xi^k_\alpha\frac{\partial}{\partial x_k}Q^{j_1\ldots j_q}_{i_1\ldots i_p} + \sum_{r=1}^{p} Q^{j_1\ldots\ldots\ldots\ldots\ldots j_q}_{i_1\ldots i_{r-1}k\,i_{r+1}\ldots i_p}\frac{\partial \xi^k_\alpha}{\partial x_{i_r}} - \sum_{r=1}^{q} Q^{j_1\ldots j_{r-1}k\,j_{r+1}\ldots j_q}_{i_1\ldots\ldots\ldots\ldots\ldots i_p}\frac{\partial \xi^{j_r}_\alpha}{\partial x_k} + W Q^{j_1\ldots j_q}_{i_1\ldots i_p}\frac{\partial \xi^k_\alpha}{\partial x_k}$$

$$= \xi^k_\alpha Q^{j_1\ldots j_q}_{i_1\ldots i_p,k} + \sum_{r=1}^{p} Q^{j_1\ldots\ldots\ldots\ldots\ldots j_q}_{i_1\ldots i_{r-1}k\,i_{r+1}\ldots i_p}\xi^k_{\alpha,i_r} - \sum_{r=1}^{q} Q^{j_1\ldots j_{r-1}k\,j_{r+1}\ldots j_q}_{i_1\ldots\ldots\ldots\ldots\ldots i_p}\xi^{j_r}_{\alpha,k}. \tag{16}$$

From the first form given for $\Delta_\alpha Q^{j_1\ldots j_q}_{i_1\ldots i_p}$ we have

$$\xi^\alpha_k \Delta_\alpha Q^{j_1\ldots j_q}_{i_1\ldots i_p} = Q^{j_1\ldots j_q}_{i_1\ldots i_p|k}.$$

The condition that a tensor should be invariant for the transformations of the group can therefore be written

$$\Delta_\alpha Q^{j_1\ldots j_q}_{i_1\ldots i_p} = 0.$$

From this we have

$$\Delta_\alpha g_{ij} = g_{kj}\xi^k_{\alpha,i} + g_{ik}\xi^k_{\alpha,j} = g_{\alpha\beta}(\xi^\beta_{j,i} + \xi^\beta_{i,j}) = 0,$$

by (12). Similarly, $\quad \Delta_\alpha g^{ij} = 0, \quad \Delta_\alpha\sqrt{g} = 0.$

It follows that the length $\{g_{ij}\xi^i\xi^j\}^{\frac{1}{2}}$ of any invariant contravariant vector ξ^i is unaltered by the transformations of the groups. Again

$$\Delta_\alpha \epsilon_{i_1\ldots i_n} = \sum_{r=1}^{n} \epsilon_{i_1\ldots i_{r-1}k\,i_{r+1}\ldots i_n}\xi^k_{\alpha,i_r}$$

$$= \epsilon_{i_1\ldots i_n}\xi^k_{\alpha,k} = 0,$$

and similarly, $\Delta_\alpha \delta^{j_1\ldots j_p}_{i_1\ldots i_p} = 0.$

56·1. **Invariant integrals.** A p-fold integral

$$\int P = \int \frac{1}{p!} P_{i_1\ldots i_p}\, dx^{i_1} \ldots dx^{i_p}$$

is said to be invariant if $\Delta_\alpha P_{i_1\ldots i_p} = 0 \qquad (\alpha = 1, \ldots, r).$

Now define $P_{\alpha_1\ldots\alpha_p} = P_{i_1\ldots i_p}\xi^{i_1}_{\alpha_1}\ldots\xi^{i_p}_{\alpha_p}.$

Then, since $\xi^i_\alpha \xi^\alpha_j = \delta^i_j,$

we have $P_{i_1\ldots i_p} = P_{\alpha_1\ldots\alpha_p}\xi^{\alpha_1}_{i_1}\ldots\xi^{\alpha_p}_{i_p},$

and $P_{\alpha_1\ldots\alpha_{r-1}\beta\,\alpha_{r+1}\ldots\alpha_p}h^\beta_\gamma = P_{\alpha_1\ldots\alpha_{r-1}\gamma\,\alpha_{r+1}\ldots\alpha_p}. \qquad (17)$

We have

$$P_{i_1\ldots i_p,k} = \left(\frac{\partial}{\partial x_k}P_{\alpha_1\ldots\alpha_p}\right)\xi^{\alpha_1}_{i_1}\ldots\xi^{\alpha_p}_{i_p}$$

$$+ \sum_{r=1}^{p} P_{\alpha_1\ldots\alpha_{r-1}\beta\,\alpha_{r+1}\ldots\alpha_p}\xi^\beta_{i_r,k}\xi^{\alpha_1}_{i_1}\ldots\xi^{\alpha_{r-1}}_{i_{r-1}}\xi^{\alpha_{r+1}}_{i_{r+1}}\ldots\xi^{\alpha_p}_{i_p}$$

$$= \left[\frac{\partial}{\partial x_k}P_{\alpha_1\ldots\alpha_p} + \sum_{r=1}^{p} P_{\alpha_1\ldots\alpha_{r-1}\beta\,\alpha_{r+1}\ldots\alpha_p}C^\gamma_{\alpha_r k}(\delta^\beta_\gamma - \tfrac{1}{2}h^\beta_\gamma)\right]\xi^{\alpha_1}_{i_1}\ldots\xi^{\alpha_p}_{i_p}$$

$$= \left[\frac{\partial}{\partial x_k}P_{\alpha_1\ldots\alpha_p} + \tfrac{1}{2}\sum_{r=1}^{p} P_{\alpha_1\ldots\alpha_{r-1}\beta\,\alpha_{r+1}\ldots x_p}C^\beta_{\alpha_r k}\right]\xi^{\alpha_1}_{i_1}\ldots\xi^{\alpha_p}_{i_p},$$

by (17). Hence,

$$\Delta_\alpha P_{i_1\ldots i_p} = \Bigg[\xi^k_\alpha\frac{\partial}{\partial x_k}P_{\alpha_1\ldots\alpha_p} + \sum_{r=1}^{p} P_{\alpha_1\ldots\alpha_{r-1}\beta\,\alpha_{r+1}\ldots\alpha_p}$$

$$\times\{\tfrac{1}{2}C^\beta_{\alpha_r\gamma}h^\gamma_\alpha + C^\beta_{\alpha_r\gamma}(\delta^\gamma_\alpha - \tfrac{1}{2}h^\gamma_\alpha)\}\Bigg]\xi^{\alpha_1}_{i_1}\ldots\xi^{\alpha_p}_{i_p}$$

$$= \left[\xi^k_\alpha\frac{\partial}{\partial x_k}P_{\alpha_1\ldots\alpha_p} + \sum_{r=1}^{p} P_{\alpha_1\ldots\alpha_{r-1}\beta\,\alpha_{r+1}\ldots\alpha_p}C^\beta_{\alpha_r\alpha}\right]\xi^{\alpha_1}_{i_1}\ldots\xi^{\alpha_p}_{i_p}.$$

Further,

$$\begin{aligned}\Delta_\alpha P_{i_1\ldots i_p}\xi^{i_1}_{\beta_1}\ldots\xi^{i_p}_{\beta_p}&=\left[\xi^k_\alpha\frac{\partial}{\partial x_k}P_{\alpha_1\ldots\alpha_p}+\sum_{r=1}^{p}P_{\alpha_1\ldots\alpha_{r-1}\beta\,\alpha_{r+1}\ldots\alpha_p}C^\beta_{\alpha_r\alpha}\right]h^{\alpha_1}_{\beta_1}\ldots h^{\alpha_p}_{\beta_p}\\&=\xi^k_\alpha\frac{\partial}{\partial x_k}P_{\beta_1\ldots\beta_p}-\sum_{r=1}^{p}P_{\beta_1\ldots\beta_{r-1}\beta\,\beta_{r+1}\ldots\beta_p}\left\{\xi^k_\alpha\frac{\partial}{\partial x_k}h^\beta_{\beta_r}-C^\beta_{\gamma\alpha}h^\gamma_{\beta_r}\right\}.\end{aligned}$$

Now, by equation (13),

$$\xi^k_\alpha\frac{\partial}{\partial x_k}h^\beta_{\beta_r}=h^\gamma_\alpha C^\beta_{\delta\gamma}h^\delta_{\beta_r}-h^\gamma_\alpha C^\delta_{\beta_r\gamma}h^\beta_\delta.$$

We may therefore write

$$\begin{aligned}&P_{\beta_1\ldots\beta_{r-1}\beta\,\beta_{r+1}\ldots\beta_p}\left\{\xi^k_\alpha\frac{\partial}{\partial x_k}h^\beta_{\beta_r}-C^\beta_{\gamma\alpha}h^\gamma_{\beta_r}\right\}\\&\quad=P_{\beta_1\ldots\beta_{r-1}\beta\,\beta_{r+1}\ldots\beta_p}h^\beta_\epsilon[C^\epsilon_{\delta\gamma}h^\delta_{\beta_r}h^\gamma_\alpha-C^\epsilon_{\delta\gamma}\delta^\delta_{\beta_r}h^\gamma_\alpha-C^\epsilon_{\delta\gamma}h^\delta_{\beta_r}\delta^\gamma_\alpha]\\&\quad=P_{\beta_1\ldots\beta_{r-1}\beta\,\beta_{r+1}\ldots\beta_p}h^\beta_\epsilon[C^\epsilon_{\delta\gamma}(h^\delta_{\beta_r}-\delta^\delta_{\beta_r})(h^\gamma_\alpha-\delta^\gamma_\alpha)-C^\epsilon_{\beta_r\alpha}]\\&\quad=-P_{\beta_1\ldots\beta_{r-1}\beta\,\beta_{r+1}\ldots\beta_p}C^\beta_{\beta_r\alpha},\end{aligned}$$

by (14) and (17). Thus the equations

$$\Delta_\alpha P_{i_1\ldots i_p}=0$$

can be replaced by

$$\xi^k_\alpha\frac{\partial}{\partial x_k}P_{\alpha_1\ldots\alpha_p}+\sum_{r=1}^{p}P_{\alpha_1\ldots\alpha_{r-1}\beta\,\alpha_{r+1}\ldots\alpha_p}C^\beta_{\alpha_r\alpha}=0. \tag{18}$$

Conversely, in order to find the invariant p-fold integrals we have to find functions $P_{\alpha_1\ldots\alpha_p}$ which satisfy (17) and (18), for $\alpha=1,\ldots,r$. In the case $r=n$, equation (17) is an identity, since

$$h^\beta_\gamma=\delta^\beta_\gamma.$$

We also observe that, on account of (18), if the tensors which define two invariant integrals are equal at any point of V they are equal at every point of V. Consequently, there cannot be more than $\binom{n}{p}$ invariant p-fold integrals which are linearly independent.

56·2. We now show that the harmonic integrals on V are invariant. Let

$$\int P = \int \frac{1}{p!} P_{i_1 \dots i_p} \, dx^{i_1} \dots dx^{i_p}$$

be a harmonic integral. Since P is closed,

$$\sum_{r=1}^{p+1} (-1)^{r-1} P_{i_1 \dots i_{r-1} i_{r+1} \dots i_{p+1}, i_r} = 0,$$

and hence $\quad P_{i_1 \dots i_p, k} = \sum_{r=1}^{p} (-1)^{r-1} P_{k i_1 \dots i_{r-1} i_{r+1} \dots i_p, i_r}.$

Therefore,

$$\begin{aligned} \Delta_\alpha P_{i_1 \dots i_p} &= \xi^k_\alpha P_{i_1 \dots i_p, k} + \sum_{r=1}^{p} P_{i_1 \dots i_{r-1} k i_{r+1} \dots i_p} \xi^k_{\alpha, i_r} \\ &= \sum_{r=1}^{p} (-1)^{r-1} \xi^k_\alpha P_{k i_1 \dots i_{r-1} i_{r+1} \dots i_p, i_r} \\ &\qquad + \sum_{r=1}^{p} (-1)^{r-1} \xi^k_{\alpha, i_r} P_{k i_1 \dots i_{r-1} i_{r+1} \dots i_p} \\ &= \sum_{r=1}^{p} (-1)^{r-1} (\xi^k_\alpha P_{k i_1 \dots i_{r-1} i_{r+1} \dots i_p})_{, i_r}. \end{aligned}$$

Hence we have $\qquad \Delta_\alpha P \sim 0.$

Since the metrical and numerical tensors are invariant, and

$$\Delta_\alpha (A^{j_1 \dots j_q}_{i_1 \dots i_p} B^{m_1 \dots m_s}_{l_1 \dots \dots l_r}) = \Delta_\alpha (A^{j_1 \dots j_q}_{i_1 \dots i_p}) B^{m_1 \dots m_s}_{l_1 \dots \dots l_r} + A^{j_1 \dots j_q}_{i_1 \dots i_p} \Delta_\alpha (B^{m_1 \dots m_s}_{l_1 \dots \dots l_r}),$$

we have $\qquad (\Delta_\alpha P)^* = \Delta_\alpha P^*,$

where P^* is the dual of P (cf. § 27·2). But, since P is harmonic,

$$P^* \to 0,$$

and therefore, by the reasoning given above,

$$\Delta_\alpha P^* \sim 0.$$

$\Delta_\alpha P$ is therefore harmonic and homologous to zero; hence

$$\Delta_\alpha P = 0,$$

and we have established the required result.

As a corollary, we have the theorem that *a manifold of n dimensions which admits a transitive, closed, semi-simple group of self-transformations has its pth Betti number not greater than* $\binom{n}{p}$.

56·3. Any invariant p-fold integral which is closed is equal to the sum of a harmonic integral and an invariant integral which is null. We now consider null invariant integrals. Let

$$\int P = \int Q_x$$

be a null invariant p-fold integral. We have

$$P_{i_1 \dots i_p} = \sum_{r=1}^{p} (-1)^{r-1} Q_{i_1 \dots i_{r-1} i_{r+1} \dots i_p, i_r},$$

and hence

$$\Delta_\alpha P_{i_1 \dots i_p} = \sum_{s=1}^{p} (-1)^{s-1} (\xi^k_\alpha P_{k i_1 \dots i_{s-1} i_{s+1} \dots i_p}), i_s.$$

Now

$$\begin{aligned} \xi^k_\alpha P_{k i_1 \dots i_{s-1} i_{s+1} \dots i_p} &= \xi^k_\alpha Q_{i_1 \dots i_{s-1} i_{s+1} \dots i_p, k} - \xi^k_\alpha (Q_{k i_2 \dots i_{s-1} i_{s+1} \dots i_p, i_1} - \dots) \\ &= \Delta_\alpha Q_{i_1 \dots i_{s-1} i_{s+1} \dots i_p} - R_{i_1 \dots i_{s-1} i_{s+1} \dots i_p}, \end{aligned}$$

where

$$R_{j_1 \dots j_{p-1}} = \sum_{r=1}^{p-1} (-1)^{r-1} (\xi^k_\alpha Q_{k j_1 \dots j_{r-1} j_{r+1} \dots j_{p-1}}), j_{r-1}.$$

Hence $$\Delta_\alpha Q_x = (\Delta_\alpha Q)_x.$$

Since P is a null form, there exists an $(n-p)$-form U satisfying

$$U^x{}_x = P,$$

by the results of Chapter III. Hence, since P is invariant,

$$\Delta_\alpha [U^x{}_x] = 0,$$

that is, $$[\Delta_\alpha U]^x{}_x = 0$$

since we have seen that the operations of forming the derived form and of forming the dual form are interchangeable with the operator Δ_α. It follows that

$$\Delta_\alpha U \rightarrow 0,$$

for otherwise

$$\int [\Delta_\alpha U]_x$$

would be a null harmonic integral. U^x is therefore an invariant $(p-1)$-form whose derivative is P. We have therefore the following results concerning invariant forms:

(i) the derived form of any invariant form is invariant;

(ii) any invariant null form is the derived form of an invariant form;

(iii) the dual of an invariant form is invariant.

56·4. We have seen that the number of linearly independent invariant p-fold integrals on V is finite. Let this number be N_p. If α_p is the number of invariant p-fold integrals no linear combination of which is closed, the number of invariant p-fold integrals which are null is α_{p-1}. Every invariant integral which is closed is the sum of an invariant null integral and a harmonic integral. Hence, if R_p is the pth Betti number of V,

$$N_p = R_p + \alpha_p + \alpha_{p-1}.$$

By (iii) of § 56·3,

$$N_p = N_{n-p},$$

and therefore

$$\alpha_p + \alpha_{p-1} = \alpha_{n-p} + \alpha_{n-p-1},$$

that is

$$\alpha_p - \alpha_{n-p-1} = -(\alpha_{p-1} - \alpha_{n-p})$$
$$= (-1)^p (\alpha_0 - \alpha_{n-1}).$$

Now an invariant function P satisfies

$$\xi^k_\alpha \frac{\partial P}{\partial x_k} = 0 \qquad (\alpha = 1, \ldots, r),$$

and therefore

$$\frac{\partial P}{\partial x_k} = 0 \qquad (k = 1, \ldots, n).$$

P is therefore a constant, that is, it is closed. Hence α_0 is zero. Now consider an invariant n-fold integral. It is the dual of an invariant 0-fold integral, and it is therefore of the form

$$\int k\sqrt{g}\,dx^1 \ldots dx^n,$$

where k is a constant. If this integral is null, k is zero, and therefore α_{n-1} must also be zero. Hence

$$\alpha_p = \alpha_{n-p-1}.$$

Let
$$\int U^i \qquad (i = 1, \ldots, \alpha_{n-p-1})$$

be the invariant $(n-p-1)$-fold integrals, no linear combination of which is closed. Then

$$\int U^i_x \qquad (i = 1, \ldots, \alpha_{n-p-1})$$

are invariant $(n-p)$-fold integrals, and

$$\int (U^i)^x \qquad (i = 1, \ldots, \alpha_{n-p-1})$$

are invariant p-fold integrals. Suppose that there is a linear combination of these last integrals which is closed, say,

$$a_i(U^i)^x \to 0.$$

Then
$$\int a_i U^i_x$$

is a harmonic integral which is null, and we conclude that

$$a_i U^i \to 0,$$

which is contrary to our hypothesis. Therefore

$$\int (U^i)^x \qquad (i = 1, \ldots, \alpha_{n-p-1})$$

are $\alpha_{n-p-1} = \alpha_p$ invariant p-fold integrals no linear combination of which is closed. It follows that any invariant p-fold integral can be written in the form

$$\int (P + W_x + U^x),$$

where P is a harmonic form, W is an invariant $(p-1)$-form, and

$$U = a_i U^i.$$

Now take
$$\int (U^i)^x = \int W^i \qquad (i = 1, \ldots, \alpha_p)$$

to be the invariant p-fold integrals, no linear combination of which is closed. By repeating the argument given above, we prove that

$$\int (W^i)^x \qquad (i = 1, \ldots, \alpha_p)$$

are $\alpha_p = \alpha_{n-p-1}$ invariant $(n-p-1)$-fold integrals, no linear combination of which is closed. Hence

$$(W^i)^x = a^i_j U^j + \psi^i \qquad (i = 1, \ldots, \alpha_p),$$

where ψ^i is a closed invariant form, and (a^i_j) is non-singular.

Let (b^i_j) be the inverse of the matrix (a^i_j). We replace U^i by $U^i + b^i_j \psi^j$. Since ψ^j is closed, this does not alter $W^i = (U^i)^x$. With this new base for the α_{n-p-1} invariant $(n-p-1)$-fold integrals, no linear combination of which is closed, we have

$$(W^i)^x = a^i_j U^j.$$

Hence
$$(W^i)^{xx} = a^i_j (U^j)^x$$
$$= a^i_j W^j,$$

and
$$(U^i)^{xx} = (W^i)^x$$
$$= a^i_j U^j.$$

The forms U^i $(i=1, \ldots, \alpha_{n-p-1})$, W^i $(i=1, \ldots, \alpha_p)$ are uniquely determined in this way as a set, but can be replaced by linear combinations of themselves with constant coefficients. It follows that the invariant factors of the matrix

$$(a^i_j - \lambda\delta^i_j)$$

are numerical invariants of the group of transformations associated with the p-fold integrals. There exists a duality relation between these sets of invariants, those corresponding to the multiplicity p being equal to those corresponding to multiplicity $n-p-1$.

Cartan [2] has shown that for spaces V with groups of a certain type, $\alpha_p = 0$ for all values of p.

57·1. The group manifold. We now consider the application of the results of §§ 56·1–56·4 to the translation groups on the group manifold M. We recall (§ 53·6) that we write the fundamental vectors of the left-hand translation group as ξ^i_α, and the fundamental vectors of the right-hand translation group as η^i_α. We have, from the results of § 53, the equations

$$\xi^j_\alpha \frac{\partial \xi^i_\beta}{\partial x_j} - \xi^j_\beta \frac{\partial \xi^i_\alpha}{\partial x_j} = C^\gamma_{\alpha\beta}\xi^i_\gamma, \tag{5}$$

$$\eta^j_\alpha \frac{\partial \eta^i_\beta}{\partial x_j} - \eta^j_\beta \frac{\partial \eta^i_\alpha}{\partial x_j} = C^\gamma_{\beta\alpha}\eta^i_\gamma, \tag{19}$$

and

$$\eta^i_\alpha \frac{\partial \eta^\alpha_k}{\partial x_j} = \xi^i_\alpha \frac{\partial \xi^\alpha_j}{\partial x_k}. \tag{20}$$

We begin by considering the left-hand translation group. Since this is a transitive, closed, semi-simple group, the theory of §§ 54–56 can be applied, and we assume the results of these paragraphs. But for the left-hand translation group we have $r = n$, and hence

$$h^\alpha_\beta = \delta^\alpha_\beta.$$

This simplifies the formulae considerably.

We first find the invariant simple integrals of the group. The necessary and sufficient condition that the integral

$$\int v_\alpha \xi_i^\alpha \, dx^i$$

should be invariant is that

$$\frac{\partial v_\alpha}{\partial x_k} + v_\beta C_{\alpha k}^\beta = 0.$$

We show that these equations are completely integrable. The condition for this is

$$v_{\alpha,ij} = v_{\alpha,ji},$$

that is,

$$-\frac{\partial v_\beta}{\partial x_j} C_{\alpha i}^\beta - \tfrac{1}{2} v_\beta C_{\alpha\gamma}^\beta C_{ij}^\gamma + \frac{\partial v_\beta}{\partial x_i} C_{\alpha j}^\beta + \tfrac{1}{2} v_\beta C_{\alpha\gamma}^\beta C_{ji}^\gamma = 0,$$

that is, $$v_\gamma C_{\beta j}^\gamma C_{\alpha i}^\beta - v_\gamma C_{\beta i}^\gamma C_{\alpha j}^\beta - v_\beta C_{\alpha\gamma}^\beta C_{ij}^\gamma = 0,$$

or $$v_\gamma [C_{\beta\mu}^\gamma C_{\alpha\lambda}^\beta + C_{\beta\lambda}^\gamma C_{\mu\alpha}^\beta + C_{\beta\alpha}^\gamma C_{\lambda\mu}^\beta] \xi_i^\lambda \xi_i^\mu = 0,$$

which is satisfied on account of (6).

We can determine the first Betti number R_1 of M by finding the number of invariant simple integrals which are closed. If

$$\int v_\alpha \xi_i^\alpha \, dx^i = \int v_i \, dx^i$$

is closed, then $$v_{i,j} = v_{j,i},$$

that is, $$\left[\frac{\partial v_\alpha}{\partial x_j} + \tfrac{1}{2} v_\beta C_{\alpha j}^\beta\right] \xi_i^\alpha - \left[\frac{\partial v_\alpha}{\partial x_i} + \tfrac{1}{2} v_\beta C_{\alpha i}^\beta\right] \xi_j^\alpha = 0,$$

which reduces to $$v_\beta C_{\lambda\mu}^\beta = 0.$$

Since $(g_{\alpha\beta}) = (C_{\alpha\mu}^\lambda C_{\lambda\beta}^\mu)$ is of rank r, the matrix $(C_{\lambda\mu}^\beta)$ of r rows and r^2 columns is of rank r, and hence it follows that $v_\beta = 0$. Hence there are no closed invariant simple integrals, and it follows that $R_1 = 0$.

It follows from this that the equations

$$\frac{\partial v_\alpha}{\partial x_k} + v_\beta C_{\alpha k}^\beta = 0$$

have a unique solution taking a given value at a certain point· We can therefore find r sets of solutions, v^λ_α, of these equations (one set corresponding to each value of λ, $\lambda = 1, \ldots, r$) which take the values $v^\lambda_\alpha = \delta^\lambda_\alpha$ at the point O representing the identity transformation. We thus obtain r invariant integrals

$$\int \zeta^\lambda_i \, dx^i = \int v^\lambda_\alpha \xi^\alpha_i \, dx^i,$$

and for the left-hand translation group of M we have the result that the number α_1 of invariant simple integrals, no linear combination of which is closed, is equal to r.

57·2. We now investigate the properties of the vectors ζ^λ_i, just defined. We define ζ^i_λ by the equation

$$\zeta^i_\lambda = g^{ij} g_{\lambda\mu} \zeta^\mu_j.$$

Then
$$\zeta^\lambda_i \zeta^i_\mu = g^{ij} g_{\mu\nu} v^\lambda_\alpha v^\nu_\beta \xi^\alpha_i \xi^\beta_j$$
$$= g_{\mu\nu} g^{\alpha\beta} v^\lambda_\alpha v^\nu_\beta.$$

Hence,
$$\frac{\partial}{\partial x_k} [\zeta^\lambda_i \zeta^i_\mu] = -g_{\mu\nu} g^{\alpha\beta} [C^\gamma_{\beta k} v^\nu_\gamma v^\lambda_\alpha + C^\gamma_{\alpha k} v^\nu_\beta v^\lambda_\gamma]$$
$$= -g_{\mu\nu} [C^{\gamma\alpha}_k + C^{\alpha\gamma}_k] v^\nu_\gamma v^\lambda_\alpha$$
$$= 0.$$

But, at the point O, which represents the identity transformation,

$$\zeta^\lambda_i \zeta^i_\mu = \xi^\lambda_i \xi^i_\mu = \delta^\lambda_\mu,$$

and therefore at every point of M we have

$$\zeta^\lambda_i \zeta^i_\mu = \delta^\lambda_\mu.$$

Hence it follows that
$$\zeta^i_\lambda v^\lambda_\alpha = \xi^i_\alpha.$$

By a similar argument we show that if

$$W^\alpha_{\lambda\mu} = v^\alpha_\beta C^\beta_{\lambda\mu} - C^\alpha_{\beta\gamma} v^\beta_\lambda v^\gamma_\mu,$$

then
$$\frac{\partial W^\alpha_{\lambda\mu}}{\partial x_k} + W^\alpha_{\lambda\nu} C^\nu_{\mu k} + W^\alpha_{\nu\mu} C^\nu_{\lambda k} = 0.$$

Since $W^{\alpha}_{\lambda\mu} = 0$ at O, it is zero everywhere, and hence

$$v^{\alpha}_{\beta} C^{\beta}_{\lambda\mu} = C^{\alpha}_{\beta\gamma} v^{\beta}_{\lambda} v^{\gamma}_{\mu}, \tag{21}$$

everywhere.

When we calculate covariant derivatives of ζ^{λ}_{i}, ζ^{i}_{λ} with respect to the symmetric connection Γ^{i}_{jk} defined by the left-hand translation group, we obtain

$$\begin{aligned}\zeta^{\lambda}_{i,j} &= -v^{\lambda}_{\gamma} C^{\gamma}_{ij} + \tfrac{1}{2} v^{\lambda}_{\gamma} C^{\gamma}_{ij} \\ &= -\tfrac{1}{2} v^{\lambda}_{\gamma} C^{\gamma}_{ij} \\ &= -\tfrac{1}{2} C^{\lambda}_{\alpha\beta} \zeta^{\alpha}_{i} \zeta^{\beta}_{j},\end{aligned}$$

by equation (21); and

$$\begin{aligned}\zeta^{i}_{\lambda,j} &= g^{ik} g_{\lambda\mu} \zeta^{\mu}_{k,j} \\ &= -\tfrac{1}{2} g^{ik} g_{\lambda\mu} C^{\mu}_{\alpha\beta} \zeta^{\alpha}_{k} \zeta^{\beta}_{j} \\ &= \tfrac{1}{2} C^{\alpha}_{\lambda\beta} \zeta^{i}_{\alpha} \zeta^{\beta}_{j}.\end{aligned}$$

Again,

$$\begin{aligned}\zeta^{i}_{\alpha} \frac{\partial \zeta^{\alpha}_{k}}{\partial x_{j}} &= \zeta^{i}_{\alpha} \zeta^{\alpha}_{k,j} + \zeta^{i}_{\alpha} \Gamma^{l}_{jk} \zeta^{\alpha}_{l} \\ &= \tfrac{1}{2} \zeta^{i}_{\alpha} v^{\alpha}_{\beta} C^{\beta}_{jk} + \Gamma^{i}_{jk} \\ &= \tfrac{1}{2} C^{i}_{jk} + \Gamma^{i}_{jk} \\ &= L^{i}_{kj} \\ &= \xi^{i}_{\alpha} \frac{\partial \xi^{\alpha}_{j}}{\partial x_{k}}.\end{aligned}$$

This last equation, and the condition that $\zeta^{i}_{\alpha} = \xi^{i}_{\alpha}$ at O, are sufficient to show that $\zeta^{i}_{\alpha} = \eta^{i}_{\alpha}$, where η^{i}_{α} is the fundamental vector corresponding to the right-hand translation group. We shall therefore write η^{i}_{α}, η^{α}_{i} for ζ^{i}_{α}, ζ^{α}_{i}, in future.

57·3. We now consider the integrals which are invariant under the left-hand translation group. Consider the integral

$$\int \frac{1}{p!} P_{i_1 \ldots i_p} \, dx^{i_1} \ldots dx^{i_p}.$$

We define

$$\overline{P}_{\alpha_1 \ldots \alpha_p} = P_{i_1 \ldots i_p} \eta^{i_1}_{\alpha_1} \ldots \eta^{i_p}_{\alpha_p}.$$

Then $$P_{i_1\ldots i_p} = \bar{P}_{\alpha_1\ldots\alpha_p}\eta^{\alpha_1}_{i_1}\ldots\eta^{\alpha_p}_{i_p}.$$

At the point O we have $P_{\alpha_1\ldots\alpha_p} = \bar{P}_{\alpha_1\ldots\alpha_p}$, but this is not true at other points. The condition that the integral is invariant is

$$\Delta_\alpha(\bar{P}_{\alpha_1\ldots\alpha_p}\eta^{\alpha_1}_{i_1}\ldots\eta^{\alpha_p}_{i_p}) = 0,$$

that is, $$\Delta_\alpha(\bar{P}_{\alpha_1\ldots\alpha_p})\,\eta^{\alpha_1}_{i_1}\ldots\eta^{\alpha_p}_{i_p} = 0,$$

that is, $$\Delta_\alpha(\bar{P}_{\alpha_1\ldots\alpha_p}) = 0.$$

Hence $\bar{P}_{\alpha_1\ldots\alpha_p}$ is a constant. Conversely, if $\bar{P}_{\alpha_1\ldots\alpha_p}$ is a constant,

$$\int\frac{1}{p\,!}\bar{P}_{\alpha_1\ldots\alpha_p}\eta^{\alpha_1}_{i_1}\ldots\eta^{\alpha_p}_{i_p}\,dx^{i_1}\ldots dx^{i_p}$$

is an invariant integral. Thus the number N_p of invariant p-fold integrals is $\binom{r}{p}$. Since, in the notation of § 56·4,

$$R_p + \alpha_p + \alpha_{p-1} = \binom{r}{p},$$

it follows that

$$\sum_{p=0}^{r}(-1)^p R_p + (-1)^r\alpha_r = \sum_{p=0}^{r}(-1)^p\binom{r}{p} = 0,$$

and since $\alpha_r = 0$, it follows that *the Euler-Poincaré invariant* $\sum_0^r(-1)^p R_p$ *of the group manifold is zero.*

57·4. Let us now consider the integrals on M which are invariant under the right-hand translation group. The fundamental vectors are now the vectors η^i_α, η^α_i. We must therefore replace the connection

$$L^i_{jk} = \xi^i_\alpha\frac{\partial\xi^\alpha_k}{\partial x_j}$$

by the connection $$L^{*i}_{jk} = \eta^i_\alpha\frac{\partial\eta^\alpha_k}{\partial x_j}$$

$$= \xi^i_\alpha\frac{\partial\xi^\alpha_j}{\partial x_k}.$$

Since $$\tfrac{1}{2}(L^{*i}_{jk}+L^{*i}_{kj}) = \tfrac{1}{2}(L^i_{jk}+L^i_{kj}) = \Gamma^i_{jk},$$
the symmetric connection is the same as for the left-hand translation group. The metrical tensor g^*_{ij} associated with the right-hand translation group is given by

$$g^*_{ij} = g_{\alpha\beta}\eta^\alpha_i\eta^\beta_j,$$

and hence at O we have $\quad g^*_{ij} = g_{ij}.$

Also, g^*_{ij} and g_{ij} satisfy the same linear differential equations

$$\frac{\partial g^*_{ij}}{\partial x_k} = g^*_{aj}\Gamma^a_{ik}+g^*_{ia}\Gamma^a_{jk},$$

and
$$\frac{\partial g_{ij}}{\partial x_k} = g_{aj}\Gamma^a_{ik}+g_{ia}\Gamma^a_{jk},$$

and from this it follows that $g_{ij} = g^*_{ij}$ everywhere. Thus the two translation groups define the same metric on the group manifold.

The condition for the invariance of a tensor $Q^{j_1\ldots j_q}_{i_1\ldots i_p}$ under the right-hand translation group is therefore

$$Q^{j_1\ldots j_q}_{i_1\ldots i_p\,||\,k} = 0.$$

From equation (10) we can deduce that

$$\xi^\alpha_{i\,||\,k} = 0,$$

and it follows as above that the invariant p-fold integrals under the right-hand translation group are the integrals

$$\int \frac{1}{p!} P_{\alpha_1\ldots\alpha_p}\,\xi^{\alpha_1}_{i_1}\ldots\xi^{\alpha_p}_{i_p}\,dx^{i_1}\ldots dx^{i_p},$$

where $P_{\alpha_1\ldots\alpha_p}$ is a constant.

57·5. The most interesting problem which arises in connection with the group manifold M is the determination of the integrals which are invariant under both the left-hand and right-hand translation groups. If such an integral is

$$\int \frac{1}{p!} P_{i_1\ldots i_p}\,dx^{i_1}\ldots dx^{i_p},$$

we have $$P_{i_1\ldots i_p\,|\,k} = 0,$$

and $$P_{i_1\ldots i_p\,||\,k} = 0.$$

Therefore $$P_{i_1\ldots i_p,\,k} = 0$$

and the conditions

$$\sum_{r=1}^{p+1} (-1)^{r-1} P_{i_1\ldots i_{r-1} i_{r+1}\ldots i_{p+1},\, i_r} = 0$$

and $$g^{rs} P_{i_1\ldots i_{p-1} r,s} = 0$$

for a harmonic integral are satisfied. Thus every integral which is invariant for both groups is a harmonic integral in the common metric. Moreover, since the metric for the two groups is the same, the harmonic integrals must be invariant for the right-hand translation group as well as for the left-hand translation group. The integrals which we are seeking are therefore just the harmonic integrals.

57·6. The condition for a harmonic integral can now be replaced by

$$P_{i_1\ldots i_p,\,k} = 0$$

and $$P_{i_1\ldots i_p\,||\,k} = 0.$$

The first of these can be written

$$\xi^k_\alpha \frac{\partial}{\partial x_k} P_{\alpha_1\ldots\alpha_p} + \tfrac{1}{2} \sum_{r=1}^{p} P_{\alpha_1\ldots\alpha_{r-1}\beta\alpha_{r+1}\ldots\alpha_p} C^\beta_{\alpha_r\alpha} = 0,$$

and the second is simply

$$\frac{\partial}{\partial x_k} P_{i_1\ldots i_p} = 0.$$

Hence the harmonic integrals are obtained by solving the equations

$$\sum_{r=1}^{p} P_{\alpha_1\ldots\alpha_{r-1}\beta\alpha_{r+1}\ldots\alpha_p} C^\beta_{\alpha_r\alpha} = 0 \qquad (\alpha = 1, \ldots, r) \qquad (22)$$

for the constants $P_{\alpha_1\ldots\alpha_p}$. We have thus reduced the problem of determining the harmonic integrals on M, and therefore the Betti numbers, to a purely algebraic problem.

We have seen that the first Betti number R_1 is zero. We now show that R_2 is also zero. The equations to determine the invariant double integrals

$$\int \tfrac{1}{2} P_{ij}\, dx^i\, dx^j = \int \tfrac{1}{2} P_{\alpha\beta}\, \xi_i^\alpha\, \xi_j^\beta\, dx^i\, dx^j$$

are
$$P_{\delta\gamma}\, C_{\beta\alpha}^\delta + P_{\beta\delta}\, C_{\gamma\alpha}^\delta = 0. \tag{23}$$

Permute α, β, γ cyclically and add the three equations obtained. We get

$$2(P_{\delta\gamma}\, C_{\beta\alpha}^\delta + P_{\delta\beta}\, C_{\alpha\gamma}^\delta + P_{\delta\alpha}\, C_{\gamma\beta}^\delta) = 0.$$

Using (23) we obtain $\quad P_{\delta\alpha}\, C_{\gamma\beta}^\delta = 0.$

Hence
$$0 = P_{\delta\alpha}\, C_{\gamma\beta}^\delta\, C_\epsilon^{\gamma\beta} = P_{\epsilon\alpha}.$$

On the other hand we have $R_3 > 0$. For,

$$\begin{aligned} C_{\lambda\beta\gamma}\, C_{\alpha\delta}^\lambda + C_{\alpha\lambda\gamma}\, C_{\beta\delta}^\lambda + C_{\alpha\beta\lambda}\, C_{\gamma\delta}^\lambda \\ &= C_{\lambda\beta\gamma}\, C_{\alpha\delta}^\lambda + g_{\alpha\epsilon}[C_{\lambda\gamma}^\epsilon\, C_{\beta\delta}^\lambda + C_{\beta\lambda}^\epsilon\, C_{\gamma\delta}^\lambda] \\ &= C_{\lambda\beta\gamma}\, C_{\alpha\delta}^\lambda + g_{\alpha\epsilon}\, C_{\delta\lambda}^\epsilon\, C_{\gamma\beta}^\lambda \qquad \text{(by (6))} \\ &= C_{\lambda\beta\gamma}\, C_{\alpha\delta}^\lambda + C_{\lambda\gamma\beta}\, C_{\alpha\delta}^\lambda = 0. \end{aligned}$$

Hence
$$\int \frac{1}{3!} C_{\lambda\mu\nu}\, \xi_i^\lambda\, \xi_j^\mu\, \xi_k^\nu\, dx^i\, dx^j\, dx^k$$

is a harmonic integral of multiplicity *three*.

We may sum up the facts already established concerning the topology of the group manifold of a closed semi-simple continuous group as follows:

(i) the group manifold is orientable;

(ii) $R_1 = R_2 = R_{r-2} = R_{r-1} = 0; \qquad R_3 = R_{r-3} \geqslant 1;$

(iii) the Euler-Poincaré invariant $\sum_0^r (-1)^p R_p = 0.$

57·7. The determination of the Betti numbers of the group manifold of a closed semi-simple continuous group has been reduced to the problem of finding the integrals on the manifold which are invariant under both the left-hand and right-hand translation groups, and these integrals, we have seen, are just the harmonic integrals associated with the metric on the manifold determined by the group. We have reduced the problem of finding these integrals to that of solving the linear equations (22). But in practice it is found that it is simpler to find the invariant integrals directly, and by finding the number of invariant p-fold integrals to determine the Betti numbers.

A semi-simple infinitesimal group is known to be the direct product of a finite number of simple groups ([3], p. 174), and when the decomposition of the group into simple groups is known the problem is simplified. Let the simple groups be $G_1, \ldots, G_k$, the dimension of G_s being r_s. We can then choose a basis for the fundamental vectors ξ^i_α so that the constants of structure $C^\gamma_{\alpha\beta}$ satisfy the conditions

$$C^\gamma_{\alpha\beta} = 0$$

unless

$$r_1 + \ldots + r_{s-1} < \alpha \leqslant r_1 + \ldots + r_s,$$

$$r_1 + \ldots + r_{s-1} < \beta \leqslant r_1 + \ldots + r_s,$$

$$r_1 + \ldots + r_{s-1} < \gamma \leqslant r_1 + \ldots + r_s,$$

for some s, and where the constants

$$C^\gamma_{\alpha\beta} \qquad (\alpha, \beta, \gamma = r_1 + \ldots + r_{s-1} + 1, \ldots, r_1 + \ldots + r_s)$$

are the constants of structure of G_s. Let $P_{\alpha_1 \ldots \alpha_p}$ be a numerical tensor (of the associated vector space) which satisfies equations (22). We consider the equations of this set for the values $\alpha = r_1 + \ldots + r_{s-1} + 1, \ldots, r_1 + \ldots + r_s$. From these equations it follows that

$$P_{\alpha_1 \ldots \alpha_p} = \Sigma A_{\alpha_1 \ldots \alpha_{p_{s-1}} \alpha_{p_s+1} \ldots \alpha_p} P^{(s)}_{\alpha_{p_{s-1}+1} \ldots \alpha_{p_s}},$$

where $P^{(s)}_{\alpha_{p_{s-1}+1} \ldots \alpha_{p_s}}$ is a numerical tensor of the associated vector space defining an invariant integral for G_s. Considering each value of s, we see that $P_{\alpha_1 \ldots \alpha_p}$ must be a linear combination of tensors of the vector space of the form

$$P^{(1)}_{\alpha_1 \ldots \alpha_{p_1}} P^{(2)}_{\alpha_{p_1+1} \ldots \alpha_{p_2}} \ldots P^{(k)}_{\alpha_{p_{k-1}+1} \ldots \alpha_p}.$$

Further, we obtain the following algorithm for the pth Betti number of the group manifold M of the semi-simple group G. The polynomial

$$\phi(t) = R_0 + R_1 t + \ldots + R_r t^r,$$

where the coefficient R_i is the ith Betti number of M, is called the Poincaré polynomial of G. Then, if $\phi_i(t)$ is the Poincaré polynomial of the simple group G_i, the Poincaré polynomial of G is

$$\phi(t) = \phi_1(t) \ldots \phi_k(t).$$

Thus in order to find the Betti numbers of the group manifold of any semi-simple group, we express it as the direct product of simple groups, and determine the Poincaré polynomials of these simple groups. This brings us to the problem of finding the Poincaré polynomials of simple groups, and we devote the remainder of this chapter to the study of these invariants for the main classes of simple groups.

58·1. **The four main classes of simple groups.** It is well known ([3], p. 180) that the closed simple groups fall into four main classes, with five isolated exceptions. The groups of these four main classes can be represented as groups of transformations

$$z'^i = a^i_j z^j$$

of the linear vector space $(z^1, \ldots, z^n)$, where the coefficients a^i_j are real or complex numbers. The four groups are represented as follows.

(i) *The unimodular group* L_n. The matrices in this case are unitary and unimodular: $\mathbf{a}'\bar{\mathbf{a}} = \mathbf{I}$, $|\mathbf{a}| = 1$. A unitary matrix $\mathbf{a}$ such that

$$|\mathbf{a} + \mathbf{I}_n| \neq 0$$

can be written, in a unique way, as

$$\mathbf{a} = (\mathbf{I}_n - \tfrac{1}{2}\boldsymbol{\alpha} - \tfrac{1}{2}i\boldsymbol{\beta})^{-1}(\mathbf{I}_n + \tfrac{1}{2}\boldsymbol{\alpha} + \tfrac{1}{2}i\boldsymbol{\beta}),$$

where $\boldsymbol{\alpha}$ is a skew-symmetric matrix of real numbers, and $\boldsymbol{\beta}$ is a symmetric matrix of real numbers. If $\mathbf{a}_1$ is a unitary matrix such that

$$|\mathbf{a}_1 + \mathbf{I}_n| = 0,$$

the unitary matrices in the neighbourhood of it can be written as

$$\mathbf{a} = \mathbf{a}_1(\mathbf{I}_n - \tfrac{1}{2}\boldsymbol{\alpha} - \tfrac{1}{2}i\boldsymbol{\beta})^{-1}(\mathbf{I}_n + \tfrac{1}{2}\boldsymbol{\alpha} + \tfrac{1}{2}i\boldsymbol{\beta}).$$

Conversely, any matrix of these forms is unitary. The condition

$$|\mathbf{a}| = 1$$

imposes one condition on the elements of $\boldsymbol{\alpha}$ and $\boldsymbol{\beta}$, and enables us to express β^n_n as a function of the other elements. Hence the dimension of the group manifold is

$$r = n^2 - 1.$$

The infinitesimal transformations of the group are given by matrices

$$\mathbf{I}_n + \boldsymbol{\alpha} + i\boldsymbol{\beta},$$

where

$$\beta^1_1 + \ldots + \beta^n_n = 0$$

on account of the unimodular condition.

(ii) *The orthogonal group* $O_{2\nu+1}$. In this case n is odd, $n = 2\nu + 1$, and the matrices $\mathbf{a}$ are real and orthogonal, and $|\mathbf{a}| = 1$. If

$$|\mathbf{a} + \mathbf{I}_n| \neq 0,$$

$\mathbf{a}$ can be written in the form

$$\mathbf{a} = (\mathbf{I}_n - \tfrac{1}{2}\boldsymbol{\alpha})^{-1}(\mathbf{I}_n + \tfrac{1}{2}\boldsymbol{\alpha}),$$

where $\boldsymbol{\alpha}$ is skew, and a similar representation can be given when

$$|\mathbf{a} + \mathbf{I}_n| = 0.$$

Thus $r = \frac{1}{2}n(n-1)$, and the infinitesimal transformations are given by matrices

$$\mathbf{I}_n + \boldsymbol{\alpha}.$$

(iii) *The orthogonal group* $O_{2\nu}$. In this case $\mathbf{a}$ is again real and orthogonal, and $|\mathbf{a}| = 1$. The properties of the group are different from those of the odd orthogonal group, but the representation is the same.

(iv) *The symplectic group* $S_{2\nu}$. In this case n is even, $n = 2\nu$, and the matrices are unitary and satisfy the equation

$$\mathbf{a}'\boldsymbol{\gamma}\mathbf{a} = \boldsymbol{\gamma},$$

where

$$\boldsymbol{\gamma} = \begin{pmatrix} \mathbf{0} & \mathbf{I}_\nu \\ -\mathbf{I}_\nu & \mathbf{0} \end{pmatrix}.$$

If

$$|\mathbf{a}+\mathbf{I}_n| \neq 0,$$

a can be expressed in the form

$$\mathbf{a} = (\mathbf{I}_n - \tfrac{1}{2}\mathbf{b})^{-1}(\mathbf{I}_n + \tfrac{1}{2}\mathbf{b}),$$

where

$$\mathbf{b} = \begin{pmatrix} \boldsymbol{\alpha}+i\boldsymbol{\gamma} & \boldsymbol{\beta}+i\boldsymbol{\delta} \\ -\boldsymbol{\beta}+i\boldsymbol{\delta} & \boldsymbol{\alpha}-i\boldsymbol{\gamma} \end{pmatrix},$$

in which $\boldsymbol{\alpha}$ is skew, and $\boldsymbol{\beta}$, $\boldsymbol{\gamma}$, $\boldsymbol{\delta}$ are symmetric, and each is a real matrix of ν rows and columns. The matrices in the neighbourhood of $\mathbf{a}_1$, where

$$|\mathbf{a}_1+\mathbf{I}_n| = 0,$$

can be represented as before. Thus $r = \tfrac{1}{2}n(n+1)$, and the infinitesimal transformations of the group are given by matrices

$$\mathbf{I}_n + \mathbf{b}.$$

We have thus obtained parametric representations of the groups. Each of them is closed. It is more symmetrical however to use superabundant coordinate systems on the group manifold, and to represent the point corresponding to the generic transformation, given by the matrix **x**, by the n^2 elements x^i_j of this matrix. The point will often be denoted by **x**, the symbol of the corresponding matrix. It is to be remembered that x^i_j is a function of r parameters $(s_1, \ldots, s_r)$.

58·2. The method of finding the invariant integrals on the group manifolds of L_n, $O_{2\nu+1}$, $O_{2\nu}$, and $S_{2\nu}$, is the same in principle for each group, and the differences are in the details. Let G be any one of the groups, M its group manifold. We first find the vectors η_i^α on M. If P_i is any covariant vector, consider the form

$$P_i ds^i$$

and make the transformation $s \to s'$, where

$$s'_i = s_i + \epsilon^\alpha \xi^i_\alpha.$$

We have

$$\begin{aligned} P_i(s')\,ds'^i &= \left[P_i(s) + \epsilon^\alpha \frac{\partial}{\partial s_j} P_i(s)\,\xi^j_\alpha\right]\left[ds^i + \epsilon^\alpha \frac{\partial \xi^i_\alpha}{\partial s_j}\,ds^j\right] \\ &= P_i(s)\,ds^i + \epsilon^\alpha \Delta_\alpha P_i ds^i + O(\epsilon^2). \end{aligned}$$

Hence if $P_i ds^i$ is invariant,

$$P_i(s')\,ds'^i = P_i(s)\,ds^i + O(\epsilon^2),$$

and conversely.

We have to find r independent 1-forms which are invariant for transformations of the left-hand translation group. Let $\mathbf{x}$ be the generic transformation of the group, and let $(X^i_j) = \mathbf{x}^{-1}$. Consider the form

$$\zeta^i_k = X^i_j dx^j_k,$$

and make the transformation $\mathbf{x} \to \mathbf{x}'$ given by

$$\mathbf{x}' = (\mathbf{I}_n + \epsilon\boldsymbol{\tau})\,\mathbf{x},$$

where $(\mathbf{I}_n + \epsilon\boldsymbol{\tau})$ is an infinitesimal transformation of the group. ($\boldsymbol{\tau}$ is, of course, independent of the elements of $\mathbf{x}$.) Since

$$X'^i_j\,dx'^j_k = X^i_j\,dx^j_k,$$

the form ζ^i_k is invariant. We have n^2 such forms, and it remains to show that r of them are independent. At the point $\mathbf{x} = \mathbf{I}_n$,

$$\mathbf{I}_n + \mathbf{x}^{-1}\frac{\partial \mathbf{x}}{\partial s_\alpha}\epsilon^\alpha = \mathbf{I}_n + \mathbf{b}$$

is an infinitesimal transformation of the group, and a (1-1) correspondence is established between the forms

$$c^k_i \zeta^i_k$$

and the infinitesimal transformations of the group. Therefore there are r independent forms ζ^i_k. Any integral which is invariant for transformations of the left-hand translation group can therefore be written in the form

$$\int \Sigma P\binom{i_1 \ldots i_p}{j_1 \ldots j_p} \zeta_{j_1}^{i_1} \times \zeta_{j_2}^{i_2} \times \ldots \times \zeta_{j_p}^{i_p}, \tag{24}$$

where the coefficients $P\binom{i_1 \ldots i_p}{j_1 \ldots j_p}$ are constants.

By reasoning similar to the above we see that if a transformation $\mathbf{x} \to \mathbf{x}'$,

$$\mathbf{x}' = \mathbf{x}\mathbf{a},$$

of the right-hand translation group is effected, ζ_k^i is transformed into

$$A_j^i \zeta_h^j a_k^h, \qquad (A_j^i) = \mathbf{a}^{-1},$$

and the condition for the invariance of (24) is therefore

$$\Sigma P\binom{i_1 \ldots i_p}{j_1 \ldots j_p} A_{h_1}^{i_1} \ldots A_{h_p}^{i_p} \zeta_{k_1}^{h_1} \times \ldots \times \zeta_{k_p}^{h_p} a_{j_1}^{k_1} \ldots a_{j_p}^{k_p}$$
$$= \Sigma P\binom{i_1 \ldots i_p}{j_1 \ldots j_p} \zeta_{j_1}^{i_1} \times \ldots \times \zeta_{j_p}^{i_p}.$$

This condition can be written as

$$\Sigma P\binom{i_1 \ldots i_p}{j_1 \ldots j_p} \begin{vmatrix} X_k^{i_1} \dfrac{\partial x_{j_1}^k}{\partial s_{\alpha_1}} & \ldots & X_k^{i_1} \dfrac{\partial x_{j_1}^k}{\partial s_{\alpha_p}} \\ \vdots & & \\ X_k^{i_p} \dfrac{\partial x_{j_p}^k}{\partial s_{\alpha_1}} & \ldots & X_k^{i_p} \dfrac{\partial x_{j_p}^k}{\partial s_{\alpha_p}} \end{vmatrix}$$
$$= P\binom{i_1 \ldots i_p}{j_1 \ldots j_p} \begin{vmatrix} A_a^{i_1} X_b^a \dfrac{\partial x_c^b}{\partial s_{\alpha_1}} a_{j_1}^c & \ldots & A_a^{i_1} X_b^a \dfrac{\partial x_c^b}{\partial s_{\alpha_p}} a_{j_1}^c \\ \vdots & & \\ A_a^{i_p} X_b^a \dfrac{\partial x_c^b}{\partial s_{\alpha_1}} a_{j_p}^c & \ldots & A_a^{i_p} X_b^a \dfrac{\partial x_c^b}{\partial s_{\alpha_p}} a_{j_p}^c \end{vmatrix}.$$

Now if $\epsilon^1, \ldots, \epsilon^r$ are r arbitrary numbers, and

$$\mathbf{b} = \left(X_k^i \frac{\partial x_j^k}{\partial s_\alpha} \epsilon^\alpha \right),$$

then $\mathbf{I}_n + \mathbf{b}$ is an infinitesimal transformation of the group, and any infinitesimal transformation of the group can be obtained in this way by suitable choice of $\epsilon^1, \ldots, \epsilon^r$. If

$\epsilon_a^1, \ldots, \epsilon_a^r$ $(a = 1, \ldots, p)$ are p sets of r numbers, we obtain p matrices $\mathbf{b}_{(a)}$, corresponding to p arbitrary infinitesimal transformations. If we multiply the above equation by

$$\begin{vmatrix} \epsilon_1^{\alpha_1} & \ldots & \epsilon_p^{\alpha_1} \\ \vdots & & \\ \epsilon_1^{\alpha_p} & \ldots & \epsilon_p^{\alpha_p} \end{vmatrix},$$

we obtain the result that

$$\Sigma P\binom{i_1 \ldots i_p}{j_1 \ldots j_p} \begin{vmatrix} b^{i_1}_{(1)j_1} & \ldots & b^{i_p}_{(1)j_p} \\ \vdots & & \\ b^{i_1}_{(p)j_1} & \ldots & b^{i_p}_{(p)j_p} \end{vmatrix} \tag{25}$$

is a polynomial in the components of the tensors $b^i_{(h)j}$ in the vector space of the transformation, which is unaltered by the transformation of the group. Conversely, we can reverse the argument to show that if (25) is invariant for all transformations of the group, where $\mathbf{b}_{(1)}, \ldots, \mathbf{b}_{(p)}$ are p linearly independent matrices such that $\mathbf{I}_n + \mathbf{b}_{(h)}$ is an infinitesimal transformation of the group, then (24) is an invariant integral. Now (25) is a polynomial in the components of the p tensors which has the properties: (i) it is linear and homogeneous in the components of each of the tensors; (ii) it is invariant under the substitutions of the group; (iii) the substitution

$$\pi = \begin{pmatrix} 1 & 2 & \ldots & p \\ \pi_1 & \pi_2 & \ldots & \pi_p \end{pmatrix}$$

performed on the tensors multiplies the polynomial by δ_π, where δ_π is $+1$ or -1 according as π is an even or an odd substitution.

58·3. Conversely, let us consider a polynomial in the components of p tensors $b^i_{(h)j}$ $(h = 1, \ldots, p)$ which are linearly independent and are such that $\mathbf{I}_n + \mathbf{b}_{(h)}$ is an infinitesimal transformation of the group. Suppose that this polynomial satisfies conditions (i) and (iii) above. It follows at once that it is of the form (25). If it also satisfies the condition (ii) it is invariant under substitutions of the group, and, as we have seen above, the corresponding integral (24) is invariant under both translation groups on the group manifold.

We therefore try to solve the problem of finding the invariant integrals by finding the polynomials in the components of the tensors $b^i_{(h)j}$ which satisfy conditions (i), (ii) and (iii). We may select the tensors $b^i_{(h)j}$ in any way we please, provided they are linearly independent and satisfy the condition that $\mathbf{I}_n + \mathbf{b}_{(h)}$ is an infinitesimal transformation of the group. In the following paragraphs we consider the groups of the four main classes separately, in each case choosing the tensors $b^i_{(h)j}$ so that we can make use of a fundamental theorem in the algebraic theory of invariants.

59·1. The unimodular group L_n. If (24) is an invariant integral for the group L_n, and we make the substitution $z \to z'$,

$$z'^i = kz^i,$$

the integral is unaltered. Hence any integral which is invariant for transformations of the unimodular group is also invariant for the transformations of the more ample group consisting of all unitary transformations. Conversely, any integral which is invariant for transformations of the more ample group is invariant for the unimodular group. We therefore have to find polynomials of the form (25) which are invariant for the transformations of the unitary group.

Since the unitary group depends on n^2 parameters, the matrices $\mathbf{b}$ are subject to no conditions. Let us take p contravariant vectors $P^i_{(h)}$ and p covariant vectors $Q^{(h)}_i$ in the n-dimensional vector space. Then we may take the p tensors to be

$$b^i_{(h)j} = P^i_{(h)} Q^{(h)}_i.$$

It is easily seen that the p tensors satisfy the condition of § 58·3. We have to find polynomials in the components of these $2p$ vectors which satisfy the conditions:

(i) they are linear and homogeneous in the components of each vector;

(ii) they are invariant under substitutions of the unitary group;

(iii) if the substitution π is made on the p tensors $P^i_{(h)}$ and the p tensors $Q_i^{(h)}$ simultaneously, the polynomial is multiplied by δ_π.

Using condition (ii), and the fundamental theorem of the theory of invariants of the unitary group [5], we know that the polynomial must be the sum of terms each of which is a product of factors of three types:

(*a*) factors of the form

$$(P_{(h)} Q^{(k)}) = P^i_{(h)} Q_i^{(k)};$$

(*b*) determinants
$$\begin{vmatrix} P^1_{(h_1)} & \dots & P^n_{(h_1)} \\ \vdots & & \\ P^1_{(h_n)} & \dots & P^n_{(h_n)} \end{vmatrix};$$

and (*c*) determinants

$$\begin{vmatrix} Q_1^{(k_1)} & \dots & Q_n^{(k_1)} \\ \vdots & & \\ Q_1^{(k_n)} & \dots & Q_n^{(k_n)} \end{vmatrix}.$$

On account of (i), each factor of the type (*b*) must be balanced by a factor of type (*c*); and since the product of these two factors is

$$\begin{vmatrix} (P_{(h_1)} Q^{(k_1)}) & \dots & (P_{(h_1)} Q^{(k_n)}) \\ \vdots & & \\ (P_{(h_n)} Q^{(k_1)}) & \dots & (P_{(h_n)} Q^{(k_n)}) \end{vmatrix},$$

the polynomial is expressible in the form

$$f = \Sigma a_\pi (P_{(1)} Q^{(\pi_1)}) \dots (P_{(p)} Q^{(\pi_p)}),$$

where the summation is over the $p\,!$ substitutions

$$\pi = \begin{pmatrix} 1 & 2 & \dots & p \\ \pi_1 & \pi_2 & \dots & \pi_p \end{pmatrix}.$$

Conversely, any polynomial of this form satisfies (i) and (ii).

59·2. We now apply condition (iii). Let πf denote the operation of replacing

$$P^i_{(1)}, \ldots, P^i_{(p)}, Q^{(1)}_i, \ldots, Q^{(p)}_i \quad \text{by} \quad P^i_{(\pi_1)}, \ldots, P^i_{(\pi_p)}, Q^{(\pi_1)}_i, \ldots, Q^{(\pi_p)}_i$$

in f, and write $R_\rho = (P_{(1)} Q^{(\rho_1)}) \ldots (P_{(p)} Q^{(\rho_p)})$.

Then, if f satisfies (iii), we have

$$\begin{aligned} f = T(f) &= \frac{1}{p!} \sum_\pi \delta_\pi \pi f \\ &= \frac{1}{p!} \sum_\pi \sum_\rho a_\rho \delta_\pi \pi R_\rho \\ &= \sum_\rho a_\rho T(R_\rho). \end{aligned}$$

Now $T(R_\rho)$ is itself a polynomial obviously satisfying (i) and (ii). If we can prove that it satisfies (iii), we shall then have merely to find the independent polynomials $T(R_\rho)$. But

$$\begin{aligned} \pi T(R_\rho) &= \pi \sum_\sigma \frac{1}{p!} \delta_\sigma \sigma R_\rho \\ &= \delta_\pi \sum_{\pi\rho} \frac{1}{p!} \delta_{\pi\sigma} (\pi\sigma) R_\rho \\ &= \delta_\pi T(R_\rho). \end{aligned}$$

Thus (iii) is also satisfied. Similarly, we prove that

$$T(\pi R_\rho) = \delta_\pi T(R_\rho).$$

Also,

$$\begin{aligned} \pi R_\rho &= (P_{(\pi_1)} Q^{(\lambda_1)}) (P_{(\pi_2)} Q^{(\lambda_2)}) \ldots (P_{(\pi_p)} Q^{(\lambda_p)}) \\ &= (P_{(1)} Q^{(\mu_1)}) (P_{(2)} Q^{(\mu_2)}) \ldots (P_{(p)} Q^{(\mu_p)}), \end{aligned}$$

where the substitutions ρ, π, λ, μ satisfy

$$\lambda = \pi\rho, \quad \lambda = \mu\pi.$$

Hence

$$\pi R_\rho = R_{\pi\rho\pi^{-1}},$$

and therefore

$$T(R_{\pi\rho\pi^{-1}}) = \delta_\pi T(R_\rho).$$

From this it follows that if there exists an odd substitution π which is permutable with ρ, $T(R_\rho)$ must be zero. We use this fact to show that, for certain substitutions ρ, $T(R_\rho)$ is zero.

We resolve the substitution

$$\rho = \begin{pmatrix} 1 & 2 & \dots & p \\ \rho_1 & \rho_2 & \dots & \rho_p \end{pmatrix}$$

into cyclic substitutions

$$\rho = (\alpha_1, \dots, \alpha_{h_1})(\alpha_{h_1+1}, \dots, \alpha_{h_1+h_2}) \dots (\alpha_{p-h_\nu+1}, \dots, \alpha_p),$$

where
$$\alpha = \begin{pmatrix} 1 & 2 & \dots & p \\ \alpha_1 & \alpha_2 & \dots & \alpha_p \end{pmatrix}$$

is a substitution on $1, \dots, p$. Suppose that, in the first place, one h, say h_a, is even. Let π be the substitution which permutes $\alpha_{h_1+\dots+h_{a-1}+1}, \dots, \alpha_{h_1+\dots+h_a}$ cyclically. Then π is odd, and it permutes with ρ. Hence

$$T(R_\rho) = 0.$$

Next, suppose that each h_a is odd, and that two of them, say h_a and h_b, are equal. Let π be the substitution which permutes

$$\alpha_{h_1+\dots+h_{a-1}+1} \text{ with } \alpha_{h_1+\dots+h_{b-1}+1}, \quad \dots, \quad \alpha_{h_1+\dots+h_a} \text{ with } \alpha_{h_1+\dots+h_b}.$$

Again π is odd, and permutes with ρ, and therefore

$$T(R_\rho) = 0.$$

Finally, take $\pi = \alpha^{-1}$; then

$$\pi\rho\pi^{-1} = (1, \dots, h_1)(h_1+1, \dots, h_1+h_2) \dots (p-h_\nu+1, \dots, p) = \rho_0$$

and
$$T(R_\rho) = \delta_\alpha T(R_{\rho_0}).$$

Any polynomial which satisfies the conditions (i), (ii) and (iii) can therefore be expressed as a sum of polynomials $T(R_{\rho_0})$, where

$$\rho_0 = (1, \dots, h_1)(h_1+1, \dots, h_1+h_2) \dots (p-h_\nu+1, \dots, p),$$

and $$h_1 < h_2 < \ldots < h_\nu, \quad h_1 + \ldots + h_\nu = p,$$
$$h_i \equiv 1 \pmod 2.$$

Now R_{ρ_0} can be written as

$$R_{\rho_0} = P_{(1)}^{i_1} Q_{i_{h_1}}^{(1)} P_{(2)}^{i_2} Q_{i_1}^{(2)} \ldots P_{(h_1)}^{i_{h_1}} Q_{i_{h_1}-1}^{(h_1)} P_{(h_1+1)}^{i_{h_1}+1} Q_{i_{h_1+h_2}}^{(h_1+1)} \ldots P_{(p)}^{i_p} Q_{i_p-1}^{(p)}$$
$$= b_{(1)i_{h_1}}^{i_1} \ldots b_{(h_1)i_{h_1}-1}^{i_{h_1}} \ldots b_{(p)i_p-1}^{i_p},$$

and therefore the polynomial (25) corresponding to $T(R_{\rho_0})$ is

$$\sum_{i_1 \ldots i_p} \begin{vmatrix} b_{(1)i_{h_1}}^{i_1} & \ldots & b_{(p)i_{h_1}}^{i_1} \\ b_{(1)i_1}^{i_2} & \ldots & b_{(p)i_1}^{i_2} \\ \vdots & & \\ b_{(1)i_p-1}^{i_p} & \ldots & b_{(p)i_p-1}^{i_p} \end{vmatrix}.$$

The corresponding integral (24) is then

$$\int \Omega_{h_1} \times \Omega_{h_2} \times \ldots \times \Omega_{h_\nu},$$

where $$\Omega_h = \zeta_{i_h}^{i_1} \times \zeta_{i_1}^{i_2} \times \ldots \times \zeta_{i_{h-1}}^{i_h}.$$

59·3. We now show that if $h > 2n-1$, Ω_h can be expressed as the sum of products

$$\Omega_{h_1} \times \Omega_{h_2} \times \ldots \times \Omega_{h_\nu},$$

where $$h_1 < h_2 < \ldots < h_\nu \leqslant 2n-1.$$

The polynomial corresponding to Ω_h is $T(R_{\rho_0})$, where ρ_0 is the cyclic substitution

$$\rho_0 = (1, 2, \ldots, h).$$

If we can show that $T(R_{\rho_0})$ can be expressed as a sum of polynomials $T(R_\rho)$ in which the substitutions ρ involve only cycles of order less than h, we can, by a simple induction, show that $T(R_{\rho_0})$ can be expressed as a sum of polynomials $T(R_\sigma)$, where σ involves only cycles of order less than $2n-1$. The result will follow immediately by constructing the corresponding integrals.

Suppose that $h = 2q-1$, where $q > n$. Then

$$0 = \begin{vmatrix} P^1_{(2)} & \dots & P^n_{(2)} & 0 & 0 & \dots & 0 \\ P^1_{(4)} & \dots & P^n_{(4)} & 0 & 0 & \dots & 0 \\ \vdots & & & & & & \\ P^1_{(2q-2)} & \dots & P^n_{(2q-2)} & 0 & 0 & \dots & 0 \\ P^1_{(2q-1)} & \dots & P^n_{(2q-1)} & 0 & 0 & \dots & 0 \end{vmatrix} \begin{vmatrix} Q^{(3)}_1 & Q^{(5)}_1 & \dots & Q^{(2q-1)}_1 & Q^{(1)}_1 \\ \vdots & & & & \\ Q^{(3)}_n & Q^{(5)}_n & \dots & Q^{(2q-1)}_n & Q^{(1)}_n \\ 0 & 0 & & 0 & 0 \\ \vdots & \vdots & & \vdots & \vdots \\ 0 & 0 & & 0 & 0 \end{vmatrix}$$

$$= \begin{vmatrix} (P_{(2)}Q^{(3)}) & \dots & (P_{(2)}Q^{(1)}) \\ \vdots & & \\ (P_{(2q-1)}Q^{(3)}) & \dots & (P_{(2q-1)}Q^{(1)}) \end{vmatrix}$$

$$= \sum_\gamma \delta_\gamma (P_{(2)}Q^{(\gamma_1)}) \dots (P_{(2q-1)}Q^{(\gamma_q)}),$$

where γ is the substitution

$$\gamma = \begin{pmatrix} 1 & 2 & 3 & \dots & 2q-2 & 2q-1 \\ \gamma_q & 2 & \gamma_1 & \dots & 2q-2 & \gamma_{q-1} \end{pmatrix}.$$

Hence

$$(P_{(1)}Q^{(2)}) \dots (P_{(2q-3)}Q^{(2q-2)}) \sum_\gamma \delta_\gamma (P_{(2)}Q^{(\gamma_1)}) \dots (P_{(2q-1)}Q^{(\gamma_q)}) = 0. \tag{26}$$

The term written explicitly is $\delta_\gamma R_\rho$, where

$$\rho = \begin{pmatrix} 1 & 2 & \dots & 2q-2 & 2q-1 \\ 2 & \gamma_1 & \dots & \gamma_{q-1} & \gamma_q \end{pmatrix}.$$

We have to consider the terms R_ρ on the left-hand side of this equation for which the substitution ρ is a cycle of order $2q-1$. For all such terms ρ is an even substitution, and since

$$\gamma^{-1}\rho = (1, 2, \dots, h)$$

is also even, $\delta_\gamma = +1$. Further, we can write the cyclic substitution ρ in the form

$$\rho = (2q-1, \gamma_q, \gamma_q+1, \lambda_2, \lambda_2+1, \dots, \lambda_{q-1}, \lambda_{q-1}+1),$$

where $\lambda_2, \dots, \lambda_{q-1}$ are odd integers. From the form of ρ, it follows that the substitution

$$\pi = \begin{pmatrix} 1 & 2 & \dots & 2q-1 \\ 2q-1 & \gamma_q & \dots & \lambda_{q-1}+1 \end{pmatrix}$$

is even.

But $$\rho_0 = \pi^{-1}\rho\pi = (1, 2, \ldots, 2q-1),$$

and therefore $T(R_\rho) = T(R_{\rho_0})$; hence, if there are k terms in (26) for which ρ is a cyclic substitution of order $2q-1$, the operation T applied to (26) gives

$$kT(R_{\rho_0}) + \Sigma T(R_\sigma) = 0,$$

where the substitutions σ only involve cycles of order less than $2q-1$. It only remains to show that k is not zero. This follows at once from the case

$$(\gamma_1, \gamma_2, \ldots, \gamma_q) = (3, 5, \ldots, 2q-1, 1).$$

59·4. We have now seen that any invariant p-fold integral on the group manifold of L_n can be written in the form

$$\Sigma a_\alpha \int \Omega_{h_1} \times \Omega_{h_2} \times \ldots \times \Omega_{h_\nu},$$

where
$$h_1 < h_2 < \ldots < h_\nu \leqslant 2n-1,$$
$$h_1 + \ldots + h_\nu = p,$$
$$h_i \equiv 1 \pmod 2.$$

Further, all these integrals are invariant. We have still to see whether the integrals

$$\int \Omega_{h_1} \times \Omega_{h_2} \times \ldots \times \Omega_{h_\nu},$$

where the suffixes h_i satisfy the above conditions, are linearly independent.

We first show that

$$\Omega_1 = \zeta_i^i = X_j^i dx_i^j = 0.$$

Since
$$\mathbf{I}_n + \left(X_j^i \frac{\partial x_k^j}{\partial s_\alpha}\epsilon^\alpha\right)$$

is an infinitesimal transformation of the group L_n, the unimodular condition gives, for all ϵ^α,

$$X_j^i \frac{\partial x_i^j}{\partial s_\alpha}\epsilon^\alpha = 0,$$

and therefore $$\Omega_1 = X^i_j dx^j_i = 0.$$

Thus, to the above conditions for the suffixes h_i, we add $h_1 > 1$.

We now show that with this new condition the integrals of multiplicity p which we obtain by taking all allowable partitions of p are independent. Since the Betti number R_r of the group manifold is *one*, there exists an invariant integral of multiplicity r, and this can only be a multiple of

$$\int \Omega_3 \times \Omega_5 \times \ldots \times \Omega_{2n-1},$$

since $3+5+\ldots+(2n-1) = n^2-1$. Hence

$$\Omega_3 \times \Omega_5 \times \ldots \times \Omega_{2n-1} \neq 0.$$

Let $$P = \Omega_{h_1} \times \Omega_{h_2} \times \ldots \times \Omega_{h_\nu}$$

be an invariant p-form. There is a unique invariant $(r-p)$-form

$$\bar{P} = \Omega_{k_1} \times \Omega_{k_2} \times \ldots \times \Omega_{k_\mu}$$

such that $h_1, \ldots, h_\nu, k_1, \ldots, k_\mu$ is a derangement of $(3, 5, \ldots, 2n-1)$,

and $$P \times \bar{P} = \Omega_1 \times \Omega_2 \times \ldots \times \Omega_{2n-1}.$$

Let $P_{(1)}, \ldots, P_{(s)}$ be the distinct forms

$$\Omega_{h_1} \times \Omega_{h_2} \times \ldots \times \Omega_{h_\nu},$$

where $$h_1 + \ldots + h_\nu = p,$$

$$1 < h_1 < \ldots < h_\nu \leqslant 2n-1,$$

$$h_i \equiv 1 \pmod 2.$$

Then $$P_{(i)} \times \bar{P}_{(j)} = 0 \qquad (i \neq j),$$

since the product involves at least one Ω_h twice; and

$$P_{(i)} \times \bar{P}_{(i)} = \Omega_3 \times \Omega_5 \times \ldots \times \Omega_{2n-1} \neq 0.$$

Suppose that the s p-forms are not independent, so that there is a relation

$$\sum_{i=1}^{s} a_i P_{(i)} = 0.$$

Then $$\sum_{i=1}^{s} a_i P_{(i)} \times \overline{P}_{(j)} = 0 \qquad (j = 1, \ldots, s),$$

and hence $$a_j = 0 \qquad (j = 1, \ldots, s),$$

that is, we have a contradiction.

We have thus obtained a basis for the invariant p-fold integrals on the group manifold M of L_n, and we can therefore calculate the Betti numbers of M. The p^d Betti number R_p is the number of partitions

$$h_1 + h_2 + \ldots + h_\nu = p$$

of p, where $$1 < h_1 < \ldots < h_\nu \leqslant 2n-1,$$

$$h_i \equiv 1 \pmod 2.$$

It follows that the Poincaré polynomial of M is

$$(1+t^3)(1+t^5)\ldots(1+t^{2n-1}).$$

60·1. The orthogonal group $O_{2\nu+1}$. The determination of the invariant integrals of the group manifolds of the other simple groups follows the same lines as for the group L_n, and the differences are in details. It will not therefore be necessary to give the arguments at such length, and we shall merely deal with the points of difference.

For orthogonal transformations in the vector space there is no difference between contravariant and covariant vectors. We may therefore write a contravariant vector $P^i_{(h)}$ as a covariant vector $P^{(h)}_i$, where

$$P^{(h)}_i = P^i_{(h)} \qquad (i = 1, \ldots, n).$$

Let $P^i_{(1)}, \ldots, P^i_{(p)}$ and $Q^{(1)}_i, \ldots, Q^{(p)}_i$ be $2p$ vectors, and let

$$b^i_{(h)j} = P^i_{(h)} Q^{(h)}_j - Q^i_{(h)} P^{(h)}_j.$$

These will be the tensors $b^i_{(h)j}$ of (25). The polynomials (25), regarded as polynomials in the components of the $2p$ vectors, satisfy the three conditions (i), (ii) and (iii) of § 58·2, but a polynomial in the components of the $2p$ vectors which satisfies these three conditions is not necessarily of the form (25). A fourth condition on the polynomials is required:

(iv) if we make an interchange of $P^i_{(h)}$, $Q^{(h)}_i$, that is, if we replace $P^i_{(h)}$ by $Q^i_{(h)}$ and $Q^{(h)}_i$ by $P^{(h)}_i$ wherever they occur, we change the sign of the polynomial. It is easily seen that when the conditions (i)–(iv) are satisfied the polynomial is of the form (25), and is invariant for the transformations of the group.

The fundamental theorem on the invariants of the orthogonal group [5] tells us that a polynomial which satisfies (i) and (ii) is a sum of terms each of which is a product of factors of the form:

(a) $$(P_{(h)}Q^{(k)}),\ (P_{(h)}P^{(k)}),\ (Q_{(h)}P^{(k)});$$

(b) determinants
$$\begin{vmatrix} P^1_{(a_1)} & \cdots & P^n_{(a_1)} \\ \vdots & & \\ P^1_{(a_\lambda)} & \cdots & P^n_{(a_\lambda)} \\ Q^1_{(b_1)} & \cdots & Q^n_{(b_1)} \\ \vdots & & \\ Q^1_{(b_\mu)} & \cdots & Q^n_{(b_\mu)} \end{vmatrix} \qquad (\lambda+\mu=n).$$

Since the product of two such determinants is expressible as the sum of products of type (a), we may suppose that each term has either one or no factor of the type (b). Suppose that there is a determinantal factor present. This factor is linear and homogeneous in the components of n vectors and therefore the factors of type (a) must be linear and homogeneous in the components of $2p-n$ vectors. But this is impossible since $2p-n$ is odd. Therefore the polynomial is expressible as the sum of products of terms of type (a).

60·2. We now take account of conditions (iii) and (iv). We consider $2^p p!$ substitutions, made up of $p!$ permutations of $P^i_{(1)}, \ldots, P^i_{(p)}$ and $Q^{(1)}_i, \ldots, Q^{(p)}_i$ simultaneously, and 2^p interchanges $(P^i_{(h)}, Q^{(h)}_i) \to (Q^i_{(h)}, P^{(h)}_i)$. A substitution is odd if it is (a) an odd permutation, or (b) a simple interchange, or (c) the product of an odd number of substitutions of type (a) and (b). If G is any polynomial in the scalar products

$$(P_{(r)}P^{(s)}),\ (Q_{(r)}Q^{(s)}),\ (P_{(r)}Q^{(s)}),$$

we now take the operation T to be defined as

$$T(G) = \frac{1}{2^p p!} \Sigma \delta_\pi \pi G,$$

summed over the $2^p p!$ substitutions, where δ_π is $+1$ or -1 according as π is an even or odd substitution. A basis for the polynomials satisfying (i)–(iv) is given by the polynomials $T(S)$, where

$$S = (P_{(\rho_1)} P^{(\rho_2)}) \dots (P_{(\rho_{2r-1})} P^{(\rho_{2r})}) (Q_{(\sigma_1)} Q^{(\sigma_2)}) \dots (Q_{(\sigma_{2r-1})} Q^{(\sigma_{2r})}) \\ \times (P_{(\tau_1)} Q^{(\tau_2)}) \dots (P_{(\tau_{k-1})} Q^{(\tau_k)}),$$

in which each vector appears once. As before, we prove that

$$\pi T(S) = \delta_\pi T(S) = T(\pi S).$$

Consider a sequence of factors

$$(P_{(\lambda_1)} P^{(\lambda_2)}) (P_{(\lambda_3)} Q^{(\lambda_4)}) \dots (Q_{(\lambda_r)} Q^{(\lambda_{r-1})})$$

of S. The interchanges

$$(P^i_{(\lambda_2)}, Q^{(\lambda_3)}_i) \to (Q^i_{(\lambda_2)}, P^{(\lambda_3)}_i), \dots, (P^i_{(\lambda_{r-1})}, Q^{(\lambda_{r-1})}_i) \to (Q^i_{(\lambda_{r-1})}, P^{(\lambda_{r-1})}_i)$$

replace this by

$$(P_{(\lambda_1)} Q^{(\lambda_2)}) (P_{(\lambda_3)} Q^{(\lambda_4)}) \dots (P_{(\lambda_{r-1})} Q^{(\lambda_r)}),$$

and do not affect the other factors of S. There thus exists a substitution π which changes S into

$$\pi S = \pm R_\rho = \pm (P_{(1)} Q^{(\rho_1)}) \dots (P_{(p)} Q^{(\rho_p)}),$$

and a basis for the polynomials is given by the polynomials $T(R_\rho)$. As in § 59·2, we show that we can confine ourselves to polynomials $T(R_\rho)$, where

$$\rho = (1, 2, \dots, h_1)(h_1+1, \dots, h_1+h_2) \dots (p-h_\nu+1, \dots, p),$$

and $\qquad h_1 < h_2 < \dots < h_\nu \leqslant 2n-1, \quad h_i \equiv 1 \pmod 2.$

60·3. We now show that if one of the numbers h_i is congruent to 1, modulo 4, then $T(R_\rho)$ is zero. Suppose, for instance, that $h_1 = 4q+1$. If the substitution π consists of the permutation

$$(2, h_1)(3, h_1-1) \dots (2q+1, 2q+2),$$

together with the interchanges

$$(P^i_{(1)}, Q^{(1)}_i) \to (Q^i_{(1)}, P^{(1)}_i), \dots (P^i_{(h_1)}, Q^{(h_1)}_i) \to (Q^i_{(h_1)}, P^{(h_1)}_i),$$

then π is an odd substitution. But π is interchangeable with ρ, and hence

$$T(R_\rho) = 0.$$

Hence a basis for the polynomials is given by the polynomials $T(R_\rho)$, where

$$\rho = (1, \dots, h_1)(h_1+1, \dots, h_1+h_2) \dots (p-h_\nu+1, \dots, p),$$

in which $$h_1 < h_2 < \dots < h_\nu \leqslant 2n-3,$$

$$h_i \equiv -1 \pmod 4.$$

The corresponding integral is

$$\int \Omega_{h_1} \times \Omega_{h_2} \times \dots \times \Omega_{h_\nu}.$$

As in § 59·4, we show that

$$\int \Omega_3 \times \Omega_7 \times \dots \times \Omega_{2n-3}$$

is the unique integral of multiplicity r. Using this, we prove the independence of the integrals as before, and show that the Poincaré polynomial for the group manifold of $O_{2\nu+1}$ is

$$(1+t^3)(1+t^7)\dots(1+t^{2n-3}) \quad (n \text{ odd}).$$

61·1. **The orthogonal group $O_{2\nu}$.** In this case we cannot eliminate the possibility of terms in our polynomial having a factor

$$\begin{vmatrix} P^1_{(a_1)} & \dots & P^n_{(a_1)} \\ \vdots & & \\ P^1_{(a_\lambda)} & \dots & P^n_{(a_\lambda)} \\ Q^1_{(b_1)} & \dots & Q^n_{(b_1)} \\ \vdots & & \\ Q^1_{(b_\mu)} & \dots & Q^n_{(b_\mu)} \end{vmatrix} \qquad (\lambda+\mu=n).$$

The difference made by the terms with a determinantal factor can be explained in the following way. Consider the improper orthogonal transformation given by the matrix

$$\mathbf{a} = \begin{pmatrix} -1 & 0 & \dots & 0 \\ 0 & 1 & & \\ \vdots & & \ddots & \\ 0 & & & 1 \end{pmatrix}.$$

This is, of course, not a transformation of the group. A term which has no determinantal factor is unaltered by the improper transformation, but a term with such a factor is changed in sign. We call the invariant integrals which are unaltered by the transformation *even* invariants, and those which are changed in sign *odd* invariants. Clearly every invariant integral is the sum of an even invariant and an odd invariant, and the process followed in the preceding paragraph serves only to find the even invariants. The even invariants are therefore compounded by addition and multiplication from the forms

$$\Omega_3, \Omega_7, \dots, \Omega_{2n-1}.$$

61·2. We now show that Ω_{2n-1} is itself compounded from the other forms. For this, it is merely necessary to show that $T(R_\rho)$, where ρ is the cyclic substitution $(1, 2, \dots, 2n-1)$, can be expressed as a polynomial $T(R_\sigma)$, where each σ only involves cycles of order less than $2n-1$.

We begin with the identity

$$\begin{vmatrix} (P_{(2)}Q^{(3)})\,(P_{(2)}Q^{(5)})\dots(P_{(2)}Q^{(2n-1)})\,(P_{(2)}Q^{(1)})\,(P_{(2)}P^{(1)}) \\ (P_{(4)}Q^{(3)}) \quad \dots \quad \dots \quad \dots \\ \vdots \\ (P_{(2n-2)}Q^{(3)})\dots \quad \dots \quad \dots \\ (P_{(2n-1)}Q^{(3)})\dots \quad \dots \quad \dots \\ (Q_{(2)}Q^{(3)}) \quad \dots \quad \dots \quad (Q_{(2)}P^{(1)}) \end{vmatrix} = 0.$$

Multiply this by $(P_{(3)}Q^{(4)})\dots(P_{(2n-3)}Q^{(2n-2)})$, and apply the operator T. If we expand the determinant in terms of the elements in the last row and last column, and make suitable interchanges in each of the terms obtained, we find, as in § 59·3, that we obtain

$$kT(R_\rho)+\Sigma T(R_\sigma) = 0,$$

where k is a positive integer, ρ is the cyclic substitution $\rho = (1, 2, \ldots, 2n-1)$ and the substitutions σ each involve cycles of order less than $2n-1$. It follows that Ω_{2n-1} can be expressed as a sum of products of $\Omega_h (h < 2n-1)$.

We can show at once that the even invariant integrals obtained in this way are independent. Indeed, in the sub-manifold of M which represents the sub-group of orthogonal transformations leaving the space $z_n = 0$ invariant, the integrals are independent; therefore they are independent in the whole manifold.

The number of even invariant integrals of multiplicity p on the group manifold M of the group $O_{2\nu}$ is therefore equal to the coefficient of t^p in the polynomial

$$(1+t^3)(1+t^7)\ldots(1+t^{2n-5}) \qquad (n \text{ even}).$$

61·3. We now come to the odd invariant integrals. If $\mathbf{a}$ is the improper transformation of § 61·1, and $\mathbf{x}$ is any transformation of the group $O_{2\nu}$, $\mathbf{axa}^{-1}$ is a transformation of the group, and in this way we define a (1-1) transformation of the group manifold M into itself. Any even invariant integral is unaltered by this transformation, while an odd invariant integral is changed in sign. Consider the point O of M corresponding to the identity transformation $\mathbf{I}_n$, and the points corresponding to the infinitesimal transformations

$$\mathbf{I}_n + \epsilon \mathbf{c}_{rs} \qquad (r = 1, \ldots, s-1;\ s = 2, \ldots, n),$$

where $\mathbf{c}_{rs}$ is the (n, n) matrix having the element in the rth row and sth column equal to $+1$, the element in the sth row and rth column equal to -1, and the remaining elements zero. Then $r+1$ points form an indicatrix for M. The transformation of M defined by the improper transformation **a** changes these points into

$$\mathbf{I}_n,\ \mathbf{I}_n - \epsilon \mathbf{c}_{1s} \qquad (s = 2, \ldots, n),$$

$$\mathbf{I}_n + \epsilon \mathbf{c}_{rs} \qquad (s > r > 1),$$

and it follows that the transformation changes the orientation of M. Now the dual of any integral is changed in sign when the orientation of the manifold is changed, and it follows from this that the dual of any even invariant integral is an odd invariant integral, and the dual of an odd invariant integral is an even invariant integral. The odd invariant integrals are therefore obtained by forming the duals of the even invariant integrals.

The number of odd invariant integrals of multiplicity p is therefore the coefficient of t^{r-p} in

$$(1+t^3)(1+t^7)\ldots(1+t^{2n-5}),$$

and this is equal to the coefficient of t^{p-n+1} in

$$t^{r-n+1}(1+t^{-3})(1+t^{-7})\ldots(1+t^{-2n+5})$$
$$= (1+t^3)\ldots(1+t^{2n-5}).$$

Hence the number of odd invariant integrals of multiplicity p is equal to the coefficient of t^p in

$$t^{n-1}(1+t^3)(1+t^7)\ldots(1+t^{2n-5}).$$

It follows that the Poincaré polynomial of the group manifold of the group $O_{2\nu}$ is

$$(1+t^3)(1+t^7)\ldots(1+t^{2n-5})(1+t^{n-1}) \qquad (n \text{ even}).$$

61·4. An odd invariant integral of multiplicity $n-1$ can be written down at once. We recall that for the orthogonal group of transformations in n-space there is no difference between covariant and contravariant vectors. We may therefore write

$$\zeta^i_j = \zeta_{ij}$$

and $$\zeta^i_j \times \zeta^j_k = Z_{ik}.$$

Then direct calculation shows that

$$(a) \qquad \Lambda = \epsilon^{i_1 \dots i_n} Z_{i_1 i_2} \times Z_{i_3 i_4} \times \dots \times Z_{i_{n-3} i_{n-2}} \times \zeta_{i_{n-1} i_n}$$

is an odd invariant form, and that

$$(b) \qquad \Omega_3 \times \Omega_7 \times \dots \times \Omega_{2n-5} \times \Lambda$$

is a non-zero invariant integral of multiplicity r. It follows that the invariant integrals on M can each be written as a sum of products of

$$\Omega_3, \dots, \Omega_{2n-5}, \Lambda,$$

in which no Ω_i or Λ appears more than once in a single term.

62. **The symplectic group** $S_{2\nu}$. This case is very similar to that of the orthogonal group $O_{2\nu+1}$. If P^i is a contravariant vector and we define

$$P_i^* = P^{\nu+i}, \; P_{\nu+i}^* = -P^i \qquad (i = 1, \dots, \nu),$$

P_i^* is a covariant vector for transformations of the symplectic group. To construct the tensors $b^i_{(h)j}$ we take $2p$ contravariant vectors $P^i_{(1)} \dots, P^i_{(p)}, Q^i_{(1)}, \dots, Q^i_{(p)}$, and write

$$b^i_{(h)j} = P^i_{(h)} Q^*_{(h)j} + P^*_{(h)j} Q^i_{(h)},$$

which fulfils the conditions required. To find the invariant polynomials (25) we have now to find polynomials in the components of the $2p$ vectors which satisfy (i) ... (iv) of § 60·1, provided we replace the simple interchange in (iv) by the interchange followed by multiplication by -1. The polynomials are expressible as the sum of terms which are products of factors:

$$(a) \qquad (P_{(r)} Q^*_{(s)}) = -(Q_{(s)} P^*_{(r)}), \quad (P_{(r)} P^*_{(s)}), \quad (Q_{(r)} Q^*_{(s)});$$

$$(b) \text{ determinants } \begin{vmatrix} P^1_{(a_1)} & \dots & P^n_{(a_1)} \\ \vdots & & \\ P^1_{(a_\lambda)} & \dots & P^n_{(a_\lambda)} \\ Q^1_{(b_1} & \dots & Q^n_{(b_1)} \\ \vdots & & \\ Q^1_{(b_\mu)} & \dots & Q^n_{(b_\mu)} \end{vmatrix} \qquad (\lambda + \mu = n).$$

Now

$$\begin{vmatrix} P^1_{(a_1)} & \cdots & P^n_{(a_1)} \\ \vdots & & \\ P^1_{(a_\lambda)} & \cdots & P^n_{(a_\lambda)} \\ Q^1_{(b_1)} & \cdots & Q^n_{(b_1)} \\ \vdots & & \\ Q^1_{(b_\mu)} & \cdots & Q^n_{(b_\mu)} \end{vmatrix} \begin{vmatrix} P^*_{(a_1)1} & \cdots & P^*_{(a_1)n} \\ \vdots & & \\ P^*_{(a_\lambda)1} & \cdots & P^*_{(a_\lambda)n} \\ Q^*_{(b_1)1} & \cdots & Q^*_{(b_1)n} \\ \vdots & & \\ Q^*_{(b_\mu)1} & \cdots & Q^*_{(b_\mu)n} \end{vmatrix} = \begin{vmatrix} (P_{(a_1)}P^*_{(a_1)}) & \cdots & (P_{(a_1)}Q^*_{(b_\mu)}) \\ \vdots & & \\ \vdots & & \\ \vdots & & \\ \vdots & & \\ (Q_{(b_\mu)}P^*_{(a_1)}) & \cdots & (Q_{(b_\mu)}Q^*_{(b_\mu)}) \end{vmatrix}$$

$$= [\Sigma \pm (P_{(a_1)}P^*_{(a_2)}) \ldots (Q_{(b_{\mu-1})}Q^*_{(b_\mu)})]^2.$$

Hence the determinant in (b) is equal to

$$\Sigma \pm (P_{(a_1)}P^*_{(a_2)}) \ldots (Q_{(b_{\mu-1})}Q^*_{(b_\mu)}).$$

Thus the determinantal factors can be eliminated.

From this point on, the reasoning is the same as in § 60. We find at once that the Poincaré polynomial of the group manifold of the symplectic group is

$$(1+t^3)(1+t^7)\ldots(1+t^{2n-1}) \qquad (n \text{ even}),$$

and that any invariant integral is expressible as the sum of products of

$$\Omega_3, \Omega_5, \ldots, \Omega_{2n-1}.$$

63. Conclusion. The method which we have given for finding the invariant integrals on the group manifolds corresponding to simple groups belonging to any of the four main classes is due to R. Brauer. Another method of obtaining the Betti numbers of these group manifolds has been given by Pontrjagin [4]. It seems possible that one of these methods could be extended to deal with the five exceptional simple groups. When this has been done, it will follow from § 56·7 that we can find the harmonic integrals on the group manifold of any closed semi-simple group once its decomposition into simple groups is known.

REFERENCES

1. E. Cartan. *La Topologie des Groupes de Lie* (Actualités Scientifiques et Industrielles, 358) (Paris), 1936.
2. E. Cartan. *Annales de la Société Polonaise de Mathématique*, 8 (1929), 181.
3. L. P. Eisenhart. *Continuous Groups of Transformations* (Princeton), 1933.
4. L. Pontrjagin. *Comptes Rendus de l'Académie des Sciences de Russie*, 1 (1935), 435.
5. H. Weyl. *The Classical Groups: their Invariants and Representations* (Princeton), 1939.

INDEX